Theodor Ackbarow

Mechanics of hierarchical alpha-helix based materials

Theodor Ackbarow

Mechanics of hierarchical alpha-helix based materials

Nanomechanical strength and fracture mechanisms

Südwestdeutscher Verlag für Hochschulschriften

Imprint
Any brand names and product names mentioned in this book are subject to trademark, brand or patent protection and are trademarks or registered trademarks of their respective holders. The use of brand names, product names, common names, trade names, product descriptions etc. even without a particular marking in this work is in no way to be construed to mean that such names may be regarded as unrestricted in respect of trademark and brand protection legislation and could thus be used by anyone.

Publisher:
Südwestdeutscher Verlag für Hochschulschriften
is a trademark of
Dodo Books Indian Ocean Ltd., member of the OmniScriptum S.R.L Publishing group
str. A.Russo 15, of. 61, Chisinau-2068, Republic of Moldova Europe
Printed at: see last page
ISBN: 978-3-8381-2100-0

Zugl. / Approved by: Paderborn, Universitaet, Diss., 2010

Copyright © Theodor Ackbarow
Copyright © 2010 Dodo Books Indian Ocean Ltd., member of the OmniScriptum S.R.L Publishing group

Abstract

Catastrophic phenomena that afflict millions of lives all have mostly one common underlying theme: the breakdown of the basic constituents leading to the failure of its overall structure and intended function. The failure and deformation of engineering materials has been studied extensively with significant impact on our world. However, the mechanisms of failure in biological systems are not well understood, thus presenting an opportunity to generate novel concepts to initiate a new paradigm of materials science. Here we undertake a systematic bottom-up analysis of the structure and properties of protein materials (PMs), illustrated by studies of intermediate filaments. We review and extend a mathematical model, which allows us to describe the mechanical strength properties of PMs in dependence of the hierarchical geometrical architecture. This model enables us to identify structure-property links and to predict the behavior of highly diverse protein structures. We validate and apply this theory in atomistic simulation studies of the fundamental fracture behavior of alpha-helix based protein domains, with and without structural defects occurring at different length and time scales. Further, we show by using a fully atomistic-informed coarse-grained multi-scale model of an alpha-helical network, that the particular architecture of IF protein networks leads to intrinsic flaw-tolerant behavior. We conclude this work by discussing the role of nanostructured hierarchies and reviewing the key findings in light of materials science concepts. Our analysis suggests that the hierarchical, nanostructured design enables PMs to unify seemingly contradicting material properties with high potential for various new bioinspired material concepts.

Zusammenfassung

Katastrophen, die Millionen von Menschenleben kosten haben meist ein zugrunde liegendes Phänomen: Das Versagen von einzelnen Bausteinen führt zum Versagen der gesamten Struktur und der vorgesehenen Funktion. Das Bruch- und Verformungsverhalten von synthetischen Materialien wurde bisher extensiv studiert und führte zur nachhaltigen Veränderung unserer Umgebung. Bisher sind jedoch die Versagensmechanismen von biologischen Systemen nicht im Detail analysiert und stellen somit eine Möglichkeit dar neue Konzepte und Paradigmen in den Materialwissenschaften zu entwickeln.

In dieser Arbeit führen wir eine systematische Analyse von apha-helix basierten Proteinmaterialien (PMs) durch. Dazu leiten wir ein mathematisches Festigkeitsmodell her, das uns Voraussagen über die Festigkeit von PMs in Abhängigkeit der geometrischen Architektur ermöglicht.

Dieses Modell wird mit atomistischen Simulationen kombiniert, um die grundlegenden Bruchmechanismen von alpha-helixbasierten (AH) Proteinen mit/ohne strukturelle Defekte auf verschiedenen Zeit- und Längsskalen zu studieren. Ebenfalls weisen wir die Fehlerrobustheit von IF-Proteinnetzwerken nach. Hierfür entwickeln wir ein weniger detailliertes und dadurch effizienteres („grobkörnigeres") Simulationsmodell.

Am Ende unserer Arbeit diskutieren wir materialwissenschaftliche und systembiologische Aspekte nanostrukturierter hierarchischer Materialien. Aus unseren Analysen kann geschlussfolgert werden, dass das hierarchische nanostrukturbasierte Design von PMs es ermöglicht, scheinbar widersprüchliche Materialeigenschaften zu verbinden und stellen somit ein hohes Potential für zahlreiche neu bioinspirierte Materialkonzepte dar.

To my parents.

Acknowledgements

I would like to express my sincere gratitude towards everyone who has supported me in the writing of this Thesis and in my research studies over the past years.

My most sincere gratitude goes to Professor Markus J. Buehler of the Massachusetts Institute of Technology (MIT, Principal Investigator of the Laboratory for Atomistic and Molecular Mechanics), who provided great excitement, continuous support and unique advice in evry respect and at any time. Due to his great efforts, we were able to publish several parts of this Thesis in peer-reviewed scientific journals and books and present these results at multiple conferences.

Moreover, I would like to thank Professor Jadran Vrabec of the Thermodynamics and Energy Technology Institute at the University of Paderborn for his support, selflessness and his trust in my work. He finally made this Thesis possible.

For their friendship, their moral and technical support, as well as the numerous fruitful discussions at work, I would like to thank my lab fellows Jérémie Bertaud, Dipanjan Sen, Sinan Keten and Rouzbeh Shahsavari. I thank Professor H. Atmanspacher, Professor E. Gekeler, M. Spiecker gen. Döhmann and Dr. Claas-Christian Wuttke for their mentorship and their belief in writing this Thesis. For their great support in difficult, time constraint situations, I would like to thank Nadine Giese and Silka Hoffmann. Many thanks to Cornelia-Anca Paulnici for her advise and support in the graphical realization of this Thesis.

The generous funding for my research project and my stay at MIT was provided by the German National Academic Foundation (Studienstiftung des deutschen Volkes), the Hamburg Foundation for research studies abroad (Hamburger Stipendienprogramm) and the Dr. Jürgen Ulderup Foundation. Additional support for this work was provided by a U.S. National Science Foundation CAREER award (grant # CMMI-0642545), a U.S. ARO grant (award # W911NF-06-1-0291), and the Solomon Buchsbaum AT&T Research Fund. Their support was gratefully received.

Finally, I am deeply indebted to my parents and my brother, who support me in an exceptional manner and always have faith in my intentions and plans. Their loving encouragement always accompanies me and this project would not have been possible without them.

Table of contents

1 **Introduction** .... - 1 -
  1.1 Nature as source of inspiration .... - 2 -
  1.2 Mechanics of materials - fracture and deformation .... - 3 -
  1.3 Outline of the Thesis .... - 3 -
2 **The protein family of intermediate filaments as model systems** .... - 5 -
  2.1 Proteins as elementary building blocks of life .... - 5 -
  2.2 Alpha-helices and coiled-coils as the elementary building blocks in IFs .... - 6 -
  2.3 The molecular architecture of vimentin IFs .... - 7 -
  2.4 The role and the mechanics of vimentin IF networks in the cytoskeleton ...... - 8 -
  2.5 The laminar networks in the nuclear membrane .... - 11 -
3 **Materials and methods: Atomistic based multi-scale simulation studies** ... - 13 -
  3.1 Atomistic and molecular modeling techniques .... - 13 -
  3.2 Simulation approach .... - 15 -
    3.2.1 CHARMM force field .... - 15 -
    3.2.2 Input data .... - 16 -
    3.2.3 Steered molecular dynamics .... - 16 -
    3.2.4 Large-scale parallelized computing .... - 17 -
    3.2.5 Data analysis and visualization methods .... - 18 -
    3.2.6 Complementary experimental methods on nanoscale .... - 19 -
  3.3 Hierarchical multi-scale modeling approach .... - 19 -
4 **Fracture mechanisms of hierarchical biological materials** .... - 21 -
  4.1 Engineering biological nanomaterials .... - 21 -
  4.2 Previous theoretical work on bond breaking dynamics .... - 23 -
  4.3 Integration of pulling speed in Bell's phenomenological model ... - 24 -
    4.3.1 Conventional Bell Model .... - 24 -
    4.3.2 Integration of pulling speed .... - 25 -
  4.4 Hierarchical Bell Model: Considering the hierarchical arrangement .... - 27 -
    4.4.1 Previous work on multiple bond cluster dynamics .... - 27 -
    4.4.2 Key assumptions for derivation .... - 28 -
    4.4.3 Derivation for a simple bond cluster .... - 30 -
    4.4.4 Derivation for hierarchically arranged bond clusters .... - 32 -
  4.5 Validation and limitations of Hierarchical Bell Model .... - 34 -
    4.5.1 Validation by direct atomistic simulations .... - 34 -
    4.5.2 Model limitations .... - 35 -
5 **Mechanics of hierarchical alpha-helical structures, from nano to macro** ... - 39 -
  5.1 Studies of primary protein structures: Point mutations .... - 39 -
    5.1.1 Protein structure .... - 40 -

- 5.1.2 Results of molecular modeling ... - 41 -
- 5.1.3 Conclusion in light of materials science and biological function ... - 43 -
- 5.2 Studies of secondary structures: Deformation and fracture in AHs ... - 44 -
  - 5.2.1 Protein structure ... - 45 -
  - 5.2.2 Mechanics at high and intermediate pulling rates ... - 46 -
  - 5.2.3 Mechanics at ultra small pulling rates ... - 50 -
  - 5.2.4 Change in deformation mode for beta-sheets ... - 54 -
  - 5.2.5 Conclusions in the light of materials science and biological function ... - 55 -
  - 5.2.6 AHs may follow Pareto principle in maximizing robustness ... - 55 -
- 5.3 Studies of tertiary structures: Defects in CCs ... - 58 -
  - 5.3.1 Protein structure ... - 58 -
  - 5.3.2 Results of molecular modeling ... - 59 -
  - 5.3.3 Conclusions in light of materials science and biological function ... - 62 -
- 5.4 Quaternary structures: concurring mechanisms of coiled-coil tetramers ... - 63 -
  - 5.4.1 Protein structure ... - 63 -
  - 5.4.2 Results of theoretical estimates ... - 64 -
  - 5.4.3 Conclusions in light of materials science and biological function ... - 67 -

6 Mechanics of multi-hierarchical systems and protein networks ... - 69 -
- 6.1 Multi-hierarchical systems ... - 69 -
  - 6.1.1 Analyzing the behavior of model systems with different hierarchies ... - 69 -
  - 6.1.2 Conclusions in light of materials science and biological function ... - 72 -
- 6.2 Flaw tolerance of alpha-helical protein networks ... - 75 -
  - 6.2.1 Modeling approach ... - 76 -
  - 6.2.2 Deformation and rupture of a AH based network ... - 79 -
  - 6.2.3 Conclusions in light of materials science and biological function ... - 86 -

7 Summary and discussion ... - 91 -
- 7.1 Summary of main results ... - 91 -
- 7.2 Nature's hierarchical tool box ... - 93 -
  - 7.2.1 System theoretical perspective on biological structures ... - 94 -
  - 7.2.2 Generic paradigm: Universality and diversity in hierarchical structures - 97 -
  - 7.2.3 Application of UDP on IFs ... - 101 -
  - 7.2.4 Significance of UDP for understanding of HBMs ... - 104 -
- 7.3 Learning from Biology, bio-inspired hierarchical structures ... - 104 -
  - 7.3.1 Interpretation of HBMs in light of materials science concepts ... - 105 -
  - 7.3.2 Robustness allows reducing safety factors ... - 105 -
  - 7.3.3 New opportunities for scientists and engineers ... - 106 -
  - 7.3.4 Impact of HBMs on other disciplines ... - 108 -
  - 7.3.5 Future challenges ... - 109 -

8 Outlook ... - 112 -

APPENDIX ... - 114 -

9 List of abbreviations and important mathematical symbols ... - 115 -

10 References ... - 117 -

Glossary

Hierarchical nanostructured materials: Material with structural features at nano-scale, typically arranged hierarchically across many length-scales. Many biological materials such as bone, spider silk or intermediate filaments (IFs) belong to this class of materials.

Proteins: Fundamental building blocks of living organisms, built from 20 basic amino acids (AAs).

Primary protein structure: Sequence of AAs/residues in a protein, building the polypeptide backbone.

Secondary protein structure: General arrangement of AAs of a polypeptide, for instance beta-sheets (BS) or alpha-helices (AHs).

Tertiary protein structure: Three-dimensional atomic coordinates of all residues in a protein, in particular including the three-dimensional, folded structure.

Quaternary protein structure: Arrangement of several proteins with different functions to a multimeric protein.

Alpha-helix (AH): Common protein secondary structure, found in many protein materials including hair, cells and hoof. Alpha-helices are hydrogen bond helical polypeptide structures stabilized by hydrogen bonds (HBs), with a diameter on the order of 1-2 nm.

Hydrogen bonds (HBs): Relatively weak inter- or intramolecular interactions, stabilizing many protein structures. HBs are crucial for the determination of the molecular properties and control many biological processes; their characteristic bond energy ranges from 2 to 8 kcal/mol.

Defect/flaw: Deviation of structural arrangement from its perfect, ideal or reference configuration, e.g. cracks.

Rupture/fracture: Sudden, typically uncontrolled loss of the equilibrium or elastic configuration of a structure or material, leading to irreversible material failure.

Robustness: Ability of a system, structure or material to tolerate flaws and defects, that is, still being capable of providing the required function.

Strength: The ability of a material or structure to withstand rupture or fracture. Strength is often compromised by defects and flaws. At nanoscale, the concept of strength must be treated using statistical theories that explicitly consider the energetic and mechanistic properties of the chemical bonds that stabilize the material or structure.

1 Introduction

The use of classes of materials has been used classify stages of civilizations, ranging from stone age thousands of years ago, to the bronze age, and possibly the silicon age in the late 20th and early 21st century. However, a systematic analysis of materials in the context of linking chemical and physical concepts, as well as the understanding and manipulation of nanostructures for engineering applications has not been achieved until quite recently. This area of nanoscience and nanotechnology led to many recent breakthroughs in

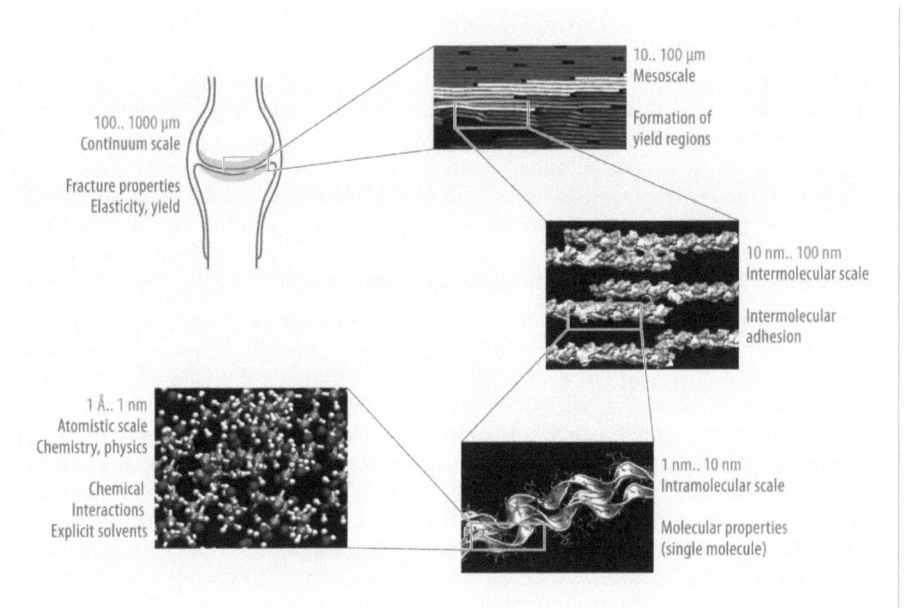

Figure 1.1: Overview over different material scales, from nano to macro, here exemplified for collagenous tissue [5-8]. Biological protein materials such as collagen, skin, bone, spider silk or cytoskeletal networks in cells feature complex, hierarchical structures. The macroscopic mechanical material behavior is controlled by the interplay of properties throughout various scales. In order to understand deformation and fracture mechanisms, it is crucial to elucidate atomistic and molecular mechanisms at each scale (examples are provided in the plot).

science, and leads to an increasing number of technological innovations.

The field of biology is probably one of the best examples for nanotechnology at work [15], since virtually all structures and materials found in biology feature nanoscale dimensions, often integrated in hierarchical patterns that link the nanoscale to larger length- and time-scales relevant for the behavior of cells, tissues and entire organisms (see, e.g. [44] for a recent overview article). Figure 1.1 shows the hierarchical structure of collagenous tissues, an example for such a hierarchically structured material. However, the understanding of how the particular length scales and structural features define the properties of protein materials and protein networks is still in its infancy, in particular the emergent properties that arise from a larger number of building blocks.

In this first Chapter, we present some of the features of biological nano-materials, the relevance of mechanical properties in this context, as well as some potential future engineering applications. This Chapter is concluded with an outline of this Thesis.

1.1 Nature as source of inspiration

Protein materials (PMs), whose structure is encoded by the DNA sequence, fulfill a variety of functions and roles in biological systems, ranging from structural support in materials such as bone, as catalysts in enzymes, for prey procurement as in claws or spider silk, or to enable sensing and communication with the outside world, such as singular GTPase proteins [15]. These multiple functions are fulfilled through materials that are synthesized at room or body temperature and which consist of chemical elements such as C (=carbon), N (=nitrogen), H (=hydrogen), O (=oxygen) or S (=sulfur), which exist on Earth in an almost unlimited amount. Both aspects underline the high level of adaptation to the environment. In contrast to that, synthetic materials consist primarily of rare elements (e.g. metals or polymers) which need to be acquired costly through deep mining or drilling (e.g. ore or oil) and which need to be treated at high temperatures in order to make them useful for technological applications.

More generally, continuous adaptation takes place in biological materials at all times, making them overall very efficient and optimized. For example, muscles or bones adapt systematically to changing environmental load by optimizing the structure according to local measures of loads (e.g. stress), leading to build-up or degradation of the structure [45]. This efficiency and dynamic adaptation of biological materials results from robust feedback loops throughout all scales, from nano to macro. These feedback loops could only evolve because biological materials "grew" under restricted environmental conditions, which required appointed properties for survival (non-fulfillment led to extermination), and due to the fact that in these materials synthesis and application appear simultaneously. This is not the case for most engineered materials up until now. These feedback loops are thus also responsible for material repair and healing after failure.

Another fascinating feature of biological materials is the realization of disparate properties within a single material, which is for example the case in bone. Here superior mechanical properties (e.g. high stiffness, high fracture resistance and high toughness) are realized through formation of a hierarchical nanocomposite composed of very soft bundles of collagen fibrils and brittle hydroxyapatite crystals. In contrast, in synthetic materials toughness and strength are disparate properties, *i.e.* hard to unify simultaneously within a single material (see Figure 6.4). The reason for this achievement in Nature may be the setup of composites with nanometer precision, by carefully combining the length and the direction of individual elements, and the hierarchical architecture of structures from nano to macro leading to a balance of different deformation mechanisms [46]. These examples illustrate that in biology, adaptation to the environment by minimizing waste was perfected, which was undoubtedly necessary for survival [47].

These are only a few of many different aspects of hierarchical biological materials and structures, rendering them a great and truly fascinating source for discovery and inspiration. To use these materials and the related structures and processes as prototypes for the development of new materials is one of the major driving forces for their study.

1.2 Mechanics of materials - fracture and deformation

As introduced above, biological materials implement multiple functions across disparate length scales. However, many of these functions are linked to mechanics at atomistic and molecular scales, since all interactions (e.g. electrostatic, van der Waals bonding etc.) communicate through forces and energies, which provide a universal language of all material processes. Thereby not only purely mechanical processes such as elastic and plastic deformation or fracture are of relevance, but also signaling processes such as mechanosensation or mechanotransduction. This makes studies of the mechanical behavior at each scale – as undertaken in this Thesis – elementary for the development of a deep understanding of the emerging structures and processes.

Up until now ceramics, metals and polymers have been in the main research focus of materials scientists. Thereby, very fundamental concepts such as dislocations in metals governed the understanding of most of the observed phenomena. However, such fundamental concepts are still missing in the relatively young field of biological nanomaterials. The approach of this Thesis is to transcend through multiple scales, from nano to macro, and to understand the underlying mechanical principles. Thus, deriving fundamental broadly applicable principles of deformation and fracture and linking them to biological functions is the main focus of this Thesis.

1.3 Outline of the Thesis

Here we provide a brief overview of the content of the entire Thesis.

Section 2 is dedicated to a thorough review of the structure (architecture as well as the elementary alpha-helical (AH) and coiled-coil (CC) building blocks), function and properties of the protein family of intermediate filaments (IFs), focused on vimentin and lamin IFs. IFs form protein networks in the cytoskeleton of eukaryotic cells, stabilize the nuclear envelope, and provide the basis for extra-cellular tissues such as hair, hoof, or nails. This protein family serves as a model system for theoretical development as well as for the validation studies reported in this Thesis. IFs exhibit many features (e.g. hierarchies, self-assembly, universal and diverse patterns, concurrent mechanisms at different length scales, etc.), which are characteristic for a vast variety of hierarchical biological materials (HBMs).

In Section 3, we briefly present molecular dynamics (MD) simulation methods, the key numerical technique used here to study the nanomechanical behavior of PMs from a fundamental bottom-up perspective.

In Section 4 we introduce a theoretical concept, which – for the first time – allows us to describe the strength of HBMs in dependence of the deformation rate and the hierarchical geometrical arrangement (*i.e.* clusters of hydrogen bonds (HBs) as well as hierarchical arrangements of these clusters in e.g. filament bundles). This theory enables us to build a structure property link, and to predict the behavior of HBM without performing experiments or simulations on that particular protein structure.

Section 5 contains the discussion of several case studies, in which we apply the theoretical concepts presented in Section 4 paired with atomistic simulation results, focusing on:

(1) The effect of amino acid (AA) point mutations on mechanics of CCs as found in laminopathies,

(2) The mechanics of AH protein structures over more than ten orders of magnitude timescales,

(3) Stutter defects in CC proteins and their implications for the mechanical behavior, as well as

(4) Concurrent deformation mechanisms of tetramers (two CC dimers).

In Section 6 we apply our theoretical findings (Section 4) and generated knowledge from MD simulations (Section 5) for studies of the mechanical behavior (strength and robustness) of multi-hierarchical model systems, and for studies of the mechanical properties (e.g. fault tolerance) of protein networks at a length scale of micrometers as they are present in the cytoskeleton as well as the cell membrane.

In Section 7 we summarize the main aspects of this Thesis and introduce a system theoretical perspective of hierarchical biological materials (HBMs), a generalized framework applicable to a wide range of biological protein structures, the universality-diversity paradigm (UDP). Then we exemplify the UDP for the case of IFs. Based on this example, we show how universality and diversity are combined through hierarchical material structures, leading to highly adapted, robust and multifunctional structures, governed through self-regulating processes. We interpret this framework in light of materials science concepts and discuss its impact on future bioinspired materials. We discuss future opportunities and new challenges that arise from a better understanding of HBMs for science, engineering and society.

The Thesis concludes in Section 8 with an outlook to future studies in this field.

2 The protein family of intermediate filaments as model systems

In this Section we provide a short presentation of biological functions, their appearance *in vivo* and previous studies of the protein family of IFs that is used as model system in this Thesis. Thereby we begin at the smallest scale and progress towards the macroscale through different levels of hierarchies.

2.1 Proteins as elementary building blocks of life

Proteins constitute the elementary building blocks of a vast variety of biological materials such as cells, spider silk or bone, where – in addition to a structural role – they govern

Cytoskeleton	A composite inside the cell consisting of three different networks: Actin filaments, microtubules and intermediate filaments. The intermediate filament network is in the focus of this Thesis. These networks connect the nucleus (nuclear membrane) with the plasma membrane and are furthermore responsible for the organization inside the cell.
Intermediate filaments (IFs)	One of the three components of the cytoskeleton; mainly responsible for the large deformation behavior of the cell.
Cross bridging proteins	Cross bridging proteins form connections inside each cytoskeletal network as well as connections to other protein systems and networks inside the cell.
Dimer	A dimer is the elementary building block of an IF fiber. This protein consists of a head domain, a tail domain and an extremely elongated coiled-coil rod. A coiled-coil is a superhelix that consists of two alpha helices that twist around each other.
Assembly	Individual IF dimers assemble systematically and hierarchically into filaments (as shown in Figure 2.3). Two dimers build a tetramer, two tetramers build an octamer and four octamers build a unit length filament (ULF). Once this level of assembly is reached, ULFs ally longitudinally into long fibers.
Residue	The primary structure of a protein consists of a sequence of amino acids. One residue is thus one amino acid in the polypeptide backbone.
Conserved structure	A structure is conserved when parts of the residue sequence are similar or do not vary at all between the different species (e.g. human and fish). For example, certain parts of the IF sequence are very similar between different species as well as inside the IF protein family (vimentin, desmin, keratin, etc.). Conserved structures often signify a particular amino acid sequence that has proven to be particularly suitable for a specific task, and has thus been kept identical during the evolutionary process.

Table 1.1: Summary of important biological terms and concepts used throughout this Thesis (descriptions adapted from [24]).

almost all processes that appear inside and outside of cells [15].

The primary structure of each protein is a polypeptide chain, consisting of a sequence of AA residues. Notably, only 20 different AAs serve as the basis in order to realize the complexity of life (albeit some AAs undergo posttranslational modifications, e.g. collagen). Each AA has its characteristic fingerprint regarding the acidity, polarity as well as hydropathy, making them overall a universal and multifunctional set of building blocks. The polypeptide chain folds into a regularly patterned secondary structure, stabilized by HBs. The most abundant secondary structures are alpha-helices (AHs) and beta-sheets (BSs). The analysis of AHs will be focus of this work. The secondary structure forms through spatial arrangement the tertiary structure, which represents the overall shape of a protein molecule. The interaction of more than one protein results in the quaternary structure. The shortly described organization of proteins illustrates the hierarchical design in protein materials (PMs). Table 1.1 summarizes the most important biological terms used in this Thesis [15].

2.2 Alpha-helices and coiled-coils as the elementary building blocks in IFs

In this Chapter we focus on AHs as well as AH CCs, the basic building blocks of IFs, an important part of the cytoskeleton of eukaryotic cells [15].

An AH is generated when a single polypeptide chain twists around on itself, stabilized by HBs made between every fourth residue, linking the O backbone atom of peptide i to the N backbone atom of peptide $i+4$ in the polypeptide chain. Consequently, at each convolution, 3.5 HBs are found in a parallel arrangement that stabilize the helical configuration [15].

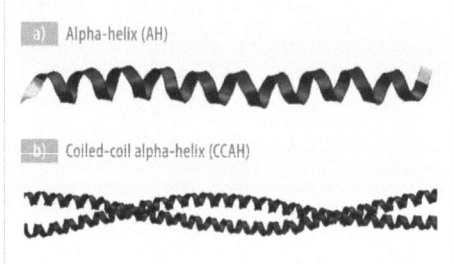

Figure 2.1: Geometry of a single AH (subplot (a)) and an AH CC molecule (subplot (b)). The CC geometry represents a super helical structure created by two AHs.

A particularly stable molecular configuration of AH based protein structures are AH CCs (CCs), which appear in approximately 10% of all proteins [48]. The CC consists of an assembly of two or more AHs in which the primary structure reveals a pronounced seven residue periodicity (abcdefg)$_n$, called heptad repeat. Within this repeat, positions "a" and "d" are preferably occupied with non-polar (hydrophobic) residues [29, 49] such as LEU, ALA, VAL or ILE. The hydrophobic residues – consequently concentrated on one side of the helix – are the reason why the proteins assemble into a CC structure. In order to avoid contact with surrounding water molecules, AHs assemble into CCs by wrapping around each other and clustering the hydrophobic side chains inside [15]. Additionally, inter-helical and intra-helical salt bridges contribute to CC thermodynamic stability [41].

The difference in molecular architecture of AHs and CCs is visualized in Figure 2.1.

Even though the heptad repeat (7/2 – seven residue repeat, whereof two residues are hydrophobic) is the most common pattern for CCs and has thus been postulated as the canonical CC, there exist other molecular structures, such as the 11/3, 15/4 or the 18/5 pattern. These structures result in three, four or five-stranded CCs [50].

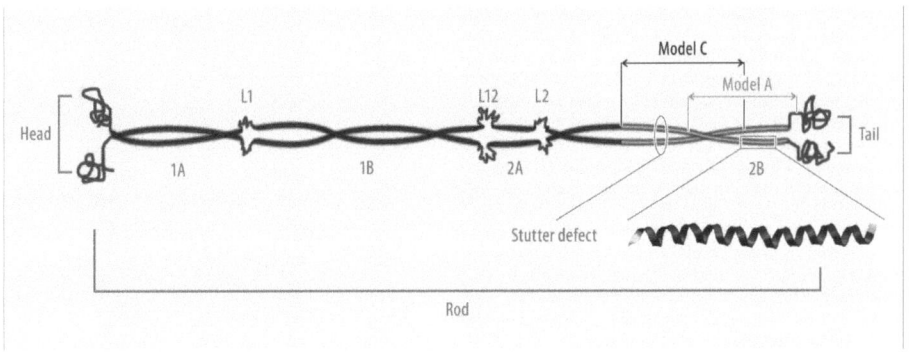

Figure 2.2: Geometry of the dimeric building block in vimentin (see, e.g. [13]). A dimer, approximately 45 nm long, is the elementary building block of (vimentin) IFs. A dimer consists of a head, tail (plotted in red) and an elongated rod domain which is divided into four AHCCs (1A, 1B, 2A, 2B) connected through linkers L1, L12, L2 (also red) [29]. Parts of the segment 1A and 2B have been crystallized so that the atomic structure is known. The stutter is located approximately in the middle of the 2B segment (indicated with a red arrow). The molecular dynamics simulations described in Section 5.3 are performed on AH CCs placed in the 2B segment (yellow). The locations of the model A and model C are shown.

2.3 The molecular architecture of vimentin IFs

Vimentin IFs are the most widely distributed type of all IFs. Vimentin proteins are typically expressed in leukocytes, blood vessel endothelial cells, some epithelial cells, and mesenchymal cells such as fibroblasts [15].

CCs are the primary building blocks of vimentin IF dimers, which are composed of a head, a tail, and an extremely elongated central rod-domain. A schematic of the vimentin dimer structure is shown in Figure 2.2. The rod-like structure is 310 residues long and consists of four CCs (1A, 1B, 2A, 2B), divided by linkers (L1, L12, L2) [29, 42, 51].

Interestingly, all helices in the IF rod domain have different lengths (the length of the 1A segment is 26 Å, 139 Å of the 1B segment, 26 Å of the 2A segment and 158 Å of the 2B segment). However, it is notable that the lengths of each of the components are absolutely conserved for all types of IFs, even in different types of cells such as leukocytes, blood vessel endothelial cells, muscle cells or neurons [15].

A variety of discontinuities, equivalent to defects from an engineering perspective, exist that interrupt the CC periodicity locally without destroying the overall molecular structure. 'Skips' are insertions of one residue into the heptad pattern [52], 'stammers' result through an insertion of three additional residues [53-55], and 'stutters' appear if four additional residues interrupt the heptad sequence [56]. Presence of a stutter results in an almost parallel run of both AHs without interrupting the CC geometry, whereas stammers lead to an over-coiling of the structure. As of to date, little is known about the biological, mechanical or physical reasons for the presence of these defects.

Thus far, only stutters have been observed in experimental analyses of the protein structure of vimentin dimers. Notably, the position of the stutter in the 2B segment is a highly conserved molecular feature (see Figure 2.2) [54]. For all known types of IFs, the

Figure 2.3: Hierarchical structure of the IF network in cells and associated characteristic length scales. Through carefully following the various steps of assembly [23, 31] it was shown that dimers associate to fibrils, which form the second level of the hierarchy. *In vivo*, these fibrils can reach a length of up to several μm and consist of 16 dimers in cross-section. The third level of hierarchy consists of three-dimensional IF-networks inside the cell, reinforcing the plasma membrane [26, 32-34]. Inside the network, IF associated proteins such as plectin generate the connection between individual IFs as well as between other cytoskeletal components (see also Figure 2.4 (a)). The characteristic loading condition of full length filaments is tensile loading. Due to this tensile load, each dimer is subject to a tensile load if the cell undergoes large deformation.

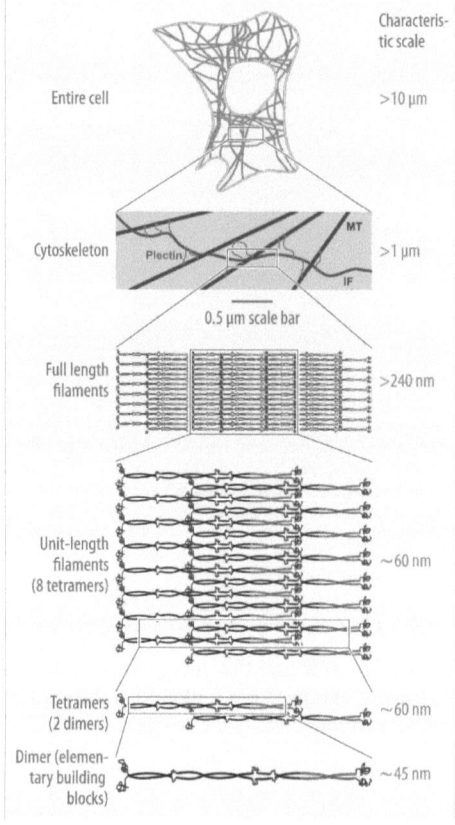

stutter is spaced precisely six heptads away from the C-terminal end of coil 2B. A detailed analysis of this irregularity will be performed in Section 5.3.

2.4 The role and the mechanics of vimentin IF networks in the cytoskeleton

Together with the globular proteins microtubules (MTs) and microfilaments (MFs), IFs are one of the three major components of the cytoskeleton in eukaryotic cells [36]. The cytoskeleton plays a critical role in determining the shape and the mechanical properties of the cell, and is vital for many additional functions including protein synthesis, cell motility as well as cell division or wound healing [33, 36, 57].

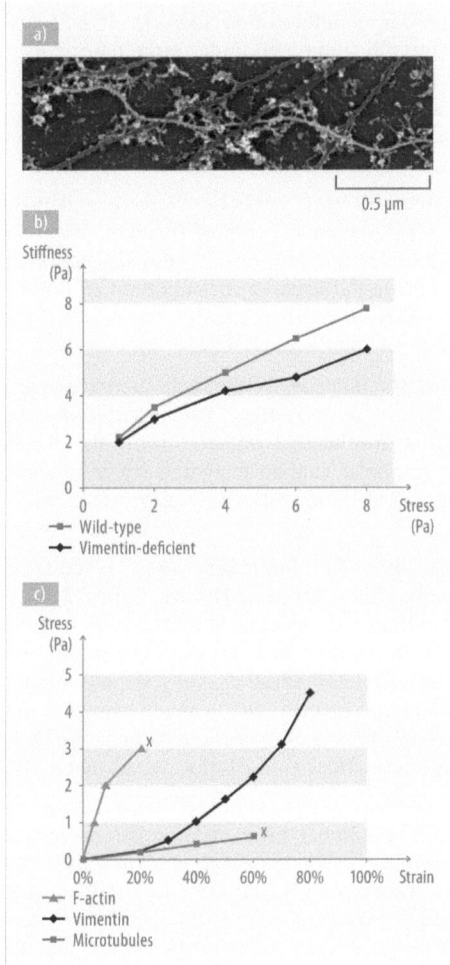

Figure 2.4: Subplot (a) depicts the architecture on the filament level, where MTs (red) and IFs (blue) are linked by plakin-type cross-bridging proteins (green). Figure taken from Alberts et al. [15]. Subplot (b) depicts the stiffness of cells as a function of stress state, comparing wild-type and vimentin deficient cells. It was shown in experiments that vimentin deficient cells are much less stiff at higher stresses than wild-type cells. These results suggest that vimentin proteins play a critical role in particular for the large-deformation elastic properties of cells. Data source: Wang et al. [36, 37]. Subplot (c) shows data from shearing experiments carried out with gels of equal weight concentration, underlining the differences in the mechanical properties of various cytoskeletal networks. In contrast to vimentin that sustains strains much larger than 80%, MT break at 60% strain, and MF break at 20% strain. Additionally, vimentin gels exhibit continuous significant strain hardening. It also corroborates the notion that due to the progressive strain hardening at large strains, IFs can be understood as "security belts" of the cell that operate after MTs and MFs have ruptured [41]. Data source: Janmey et al. [42].

Hierarchical structure of IFs

Like many other biological materials, IFs are hierarchical structures with highly specific features at nanoscale. Figure 2.3 depicts details of the hierarchical structure of the IF network, summarizing the structural features observed throughout several length scales. Vimentin IF dimers are the elementary building blocks of IFs. Through carefully following the various steps of assembly [23, 31] it was shown that dimers associate to fibrils. Fibrils build the second level of the hierarchy. *In vivo*, these fibrils can reach a length of up to several μm and consist of 16 dimers in cross-section. The third level of hierarchy consists of three-dimensional IF-networks inside the cell, reinforcing the plasma membrane [32-34]. Inside the network, IF associated proteins such as plectin generate the connection between individual filaments (see Figure 2.3 level cytoskeleton, and also Figure 2.4 (a)).

The IF networks are connected with other cellular networks, as well as with the extra cellular matrix at the plasma membrane [15]. This architecture guarantees that tensile and shear loads applied to the tissue can be carried by IF networks.

Mechanical functions of IFs

Here we focus exclusively on the mechanical role of IFs. Biologically, vimentin IFs (and also lamin IFs) are primarily associated with carrying passive loads applied to cells as well as mechanotransduction, in particular at large deformation [32-34]. It has been hypothesized that IFs are critical to provide strength to the cell under large deformation, and to absorb large amounts of energy upon a certain load by unfolding (see Figure 2.4 (b)) [22, 58]. Figure 2.4 (c) shows the stress strain behavior of vimentin IFs compared to actin filaments and microtubules. This comparison underlines their passive mechanical role as the security belts of the cell.

Under deformation of the entire IF network in a cell each vimentin filament undergoes tensile deformation. On the individual protein level, the tensile load of filaments is carried by individual dimers. Therefore, a detailed understanding of CC dimers and their mechanical properties under small and large tensile deformation is important to provide insight into the function and mechanisms of vimentin filaments and networks. We will undertake detailed studies to address this issue in Chapter 5.2 and 6.3.

Further, since IF filaments span from the cell's nucleus to the cell membrane and therefore interact with IF networks of other cells (via desmosomes), suggests that IFs play an important role in sensing and transmitting mechanical signals from the plasma membrane to the nucleus, where a specific response can be triggered by mechanical stimulation [59, 60]. Thus, it has been suggested that IF networks may play a vital role in mechanotransduction [32-34, 61, 62].

Plakin-type cross-bridging proteins, also known as cytolinkers (e.g. plectins or desmoplakins) link all three cytoskeletal networks (MTs, MFs and IFs, see Figure 2.4 (a)). These proteins attach the IFs to MTs, MFs or adhesion complexes of membranes (e.g. the cell membrane or the nuclear membrane [63]). In contrast to MTs and MFs, IFs do not participate in the dynamic functions of the cytoskeleton. Further, they do not support active transport of motor proteins such as myosin and kinesin, due to the missing polarity in the protein structure [29]. They do not participate in any cell movement [15]. These examples further underline the specific static-mechanical role of IFs, which are in the focus of this Thesis.

Experiments have shown that IFs exhibit a highly nonlinear stress-strain relationship with a high resistance against rupture at large deformation. For instance, as shown in Figure 2.4 (b) in shear tests vimentin deficient cells were shown to be 40% less stiff at large strains compared with wild-type cells, while their elastic properties do not change much under small deformation [36]. These experiments strongly support the notion that the biomechanical significance of vimentin IFs lies in the large-deformation regime, while the flexibility of IF at small strains and loading rates (compared to the properties of MFs), enables a lower mechanical resistance during cell movement, underlining the mechanical multifunctionality of IF networks.

In contrast to IFs, other components of the cytoskeleton network have been shown to rupture at much lower strains, as shown in Figure 2.4 (c). MFs rupture at low strains but large forces, and MTs break at moderately large strains, but small forces. This provides additional support for the significance of vimentin as the 'security belt' of the cell. Furthermore the different mechanical properties of the cytoskeletal networks clearly indicate that the cytoskeleton is a composite with a range of mechanical properties, which cannot be achieved by a polymer network composed out of a single type of polymer.

Observations in rupture experiments of single IFs have shown a dramatic change in filament diameter, which remained unchanged for several hours after rupture appeared [64, 65]. This is an indication for a profound change in the molecular architecture under large deformation. This provided some evidence that the mechanical properties of IFs mainly depend on the nano-mechanical, molecular properties of the CC dimer [64]. However, no direct experimental, simulation or theoretical proof has been reported thus far.

Finally, the mechanical role of IFs is particularly evident in diseases in which the loss of mechanical function and integrity of various tissues is associated with IF protein mutations [66, 67]. It was shown that mutations in keratin IFs reduce the ability of these IF networks to bundle and to resist large deformation [36]. Furthermore, it has been suggested that point mutations lead to the aggregation of the cytoskeleton and extensive cell fragility in epidermis, heart and skeletal muscle after they are exposed to mechanical strain [68]. These examples clearly illustrate the significance of the mechanical properties of IF proteins for biological processes.

In this Thesis we focus on the mechanical properties of individual AHs, CC-segments as well as interdimer interactions. We further analyze the mechanics of whole protein network, as they appear in cells. Figure 5.1 depicts an overview on the studied length scales and the key topics that are addressed in this Thesis.

2.5 The laminar networks in the nuclear membrane

Lamin IFs are a members of the IF protein family, which are located in contrast to vimentin IFs inside the cell's nuclear membrane and the nucleus. There are primarily two types of lamins, lamin A/C and lamin B1/B2, each of which is encoded by different genes. A lamin dimer has a similar protein structure as vimentin. The main difference appears in the tail domain, where lamin dimers have an additional globular domain consisting of a beta sheet structure (see Figure 5.2 for details) [34]. Dimers assemble into complex hierarchical structures, built out of three hierarchical levels: The next higher level is the assembly of dimers to filaments through a head to tail interaction of individual proteins [32, 69]. These filaments form the third hierarchical level, a dense party mesh-like network called lamina [28].

Diseases related to mutations in lamins

In contrast to vimentin, laminar proteins are much less understood at this point. However, in recent years, lamin has received increasing attention in the scientific community, because more than 200 mutations have been discovered in the lamin A/C gene [70] that are directly associated with a class of diseases referred to as laminopathies. Diseases that have been linked to lamin mutations include muscle dystrophy, lipodystrophy, neuropathy, progeria, as well as cancer [71-73]. Several studies with mutations related to these diseases have provided evidence that the lamina network is significant for maintaining the structural and mechanical integrity of the nucleus [74-77], leading to two hypotheses about its biological function and role: The structure hypothesis suggests that mutations in lamin cause fragility of the nuclear membrane and eventually facilitates damage of the nuclear membrane leading to cell death. The gene expression hypothesis suggests that mutations in lamins interrupt the normal transcription of mechanically activated proteins. However, recent studies have shown that lamin deficiency can cause both structural and transcriptional abnormalities in cells, suggesting that both hypotheses may be valid [75]. Earlier studies have already shown a significant influence of the mechanical properties of the lamin network in laminopathies [60].

Even though it has been suggested that mutations related to muscle dystrophy lead to an abnormality in the mechanical properties of the nucleus (e.g. heart failure at young age) [78, 79], up to date only limited knowledge exists about the hierarchical scale at which a single mutation causes an abnormality in the mechanical properties. In particular, it remains unclear if the mutations cause changes in the structure and properties on the individual dimer level, the filament level or the network level. Clarification of this issue could help us to better understand the origin of these diseases and help to develop more directed medical treatments.

In Chapter 5.1 of this Thesis we focus on individual point mutations in the lamin 2B segment, which are known to cause muscle dystrophies, thereby we investigate the effects of mutations on the mechanical properties of the protein structure.

3 Materials and methods: Atomistic based multi-scale simulation studies

In this Section we introduce the atomistic modeling methods used for most studies presented throughout this Thesis.

3.1 Atomistic and molecular modeling techniques

Atomistic molecular dynamics (MD) is an useful tool for elucidating the atomistic mechanisms that control deformation and rupture of chemical bonds at nano-scale, and to relate this information to macroscopic fracture phenomena (see, e.g. general review articles [80, 81], and recent articles on large-scale MD simulation of brittle fracture mechanisms [82-86]). The basic concept behind atomistic simulation via MD is to calculate the dynamical trajectory of each atom in the material, by considering their atomic interaction potentials, by solving each atom's equation of motion according to Newton's law $F = m \cdot a$. Numerical integration of this equation by considering proper interatomic potentials enables one to simulate a large ensemble of atoms that represents a material volume, albeit typically limited to several hundred nanoseconds of time intervals. The availability of such potentials for a specific material is often a limiting factor for the applicability of this method. Computational power, even if growing exponentially (see Figure 3.3) is the other current limitation.

Classical molecular dynamics generates trajectories of a large number of N particles, interacting with a specific interatomic potential, leading to positions $r_i(t)$, velocities $v_i(t)$ and accelerations $a_i(t)$. It can be considered as an alternative approach to methods like Monte-Carlo, with the difference that MD provides full dynamical information and deterministic trajectories. This particularity is of special interest, as among other deformation dynamics, bond breaking and the related unfolding mechanisms are in focus of this thesis. It is emphasized that Monte-Carlo schemes provide certain advantages as well; however, this point will not be discussed further here as all simulation studies presented here are carried out with a MD approach.

We now introduce the basic mathematical concept, which is the backbone of MD simulations. The total energy of the system is written as the sum of kinetic energy (K) and potential energy (U),

$$E = K + U \tag{3.1}$$

where the kinetic energy is

$$K = \frac{1}{2}\sum_{j=1}^{N} m_j v_j^2, \tag{3.2}$$

and the potential energy is a function of the atomic distances r_{ij} (two-body interactions are assumed),

$$U = U(r_{ij}), \tag{3.3}$$

with a properly defined potential energy surface $U(r_{ij})$. The numerical problem to be solved is a system of coupled 2nd order nonlinear differential equations:

$$m_j \frac{d^2 r_j}{dt^2} = -\nabla_{r_j} U(r_{ij}) \quad j = 1....N, \quad (3.4)$$

which can only be solved numerically for more than two particles, $N > 2$. Typically, MD is based on updating schemes that yield new positions from the old positions, velocities and the current accelerations of particles:

$$r_j(t_0 + \Delta t) = -r_j(t_0 - \Delta t) + 2r_j(t_0) + a_j(t_0)(\Delta t)^2 + ... \quad (3.5)$$

The forces and accelerations are related by $a_j = f_j / m_j$. The forces are obtained from the potential energy surface – sometimes also called force field – as

$$f_j = m_j \frac{d^2 r_j}{dt^2} = -\nabla_{r_j} U(r_j) \quad j = 1..N. \quad (3.6)$$

This technique can also be used not only for predicting the dynamics of single atoms, but also groups of atoms as in the case of coarse-grained mesoscale approaches. Provided that appropriate interatomic potentials are available, MD is capable of directly simulating a variety of materials phenomena, for instance the response of an atomic crystal lattice to applied loading under the presence of a crack-like defect, or the deformation mechanisms of biomolecules including nucleic acids and proteins.

One of the strengths of atomistic methods is its very fundamental viewpoint of materials phenomena. The only physical law that is put into the simulations is Newton's law and a definition of how atoms interact with each other. Despite this very simple basis, very complex phenomena can be simulated. Unlike many continuum mechanics approaches primarily used in engineering practice today, atomistic techniques require no *a priori* assumption of the defect dynamics. Once the atomic interactions are chosen, the complete

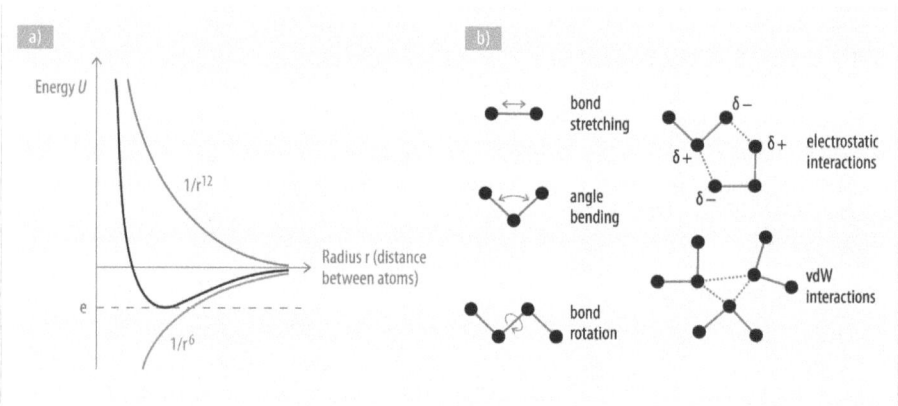

Figure 3.1: subplot (a) depicts a widely applied potential for non-bonded interactions, the 12-6 Lennard Jones Force Field. Subplot (b) Shows the individual energy contributions due to bond stretching, bond bending, bond rotation as well as electrostatic and vdW interactions. The combination of these terms constitutes the entire energy landscape of interatomic and intermolecular interactions, as given in equation (3.7). Figure adopted from [12].

material behavior is determined. Choosing appropriate models for interatomic interactions, however, provides a rather challenging and crucial step that remains subject of a very active discussion in the scientific community. A variety of different interatomic potentials are used in the studies of biological materials at different scales, and different types of protein structures require the use of different atomistic models mostly derived from QM simulations such as Density Functional Theory (DFT) (see Figure 3.4). A drawback of atomistic simulations is the difficulty of analyzing results and the large computational resources necessary to perform the simulations. Due to computational limitations, MD simulations are restricted with respect to the time scales that can be reached, limiting overall time spans in such studies to tens of nanoseconds, or in very long simulation studies to fractions of microseconds. Therefore, many MD simulation results of dynamically stretching tropocollagen molecules or IF protein domains, for instance, have been carried out at large deformation rates, exceeding several m/s, which are much higher than deformation rates that appear *in vivo* (nm/s). A solution of how this time gap can be solved will be presented in Chapter 5.2.

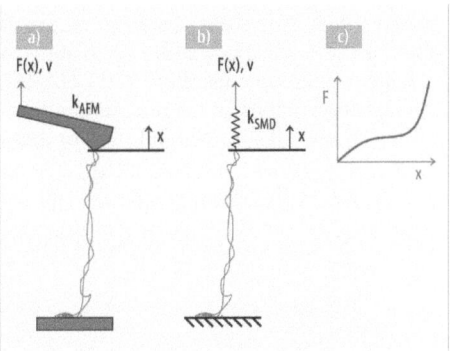

Figure 3.2: Single molecule pulling experiments, carried out on a CC protein structure. Subplot (a) depicts an experimental setup based on Atomic Force Microscopy (AFM), and subplot (b) depicts a Steered Molecular Dynamics (SMD) analogue. In the SMD approach, the end of the molecule is slowly pulled with a pulling velocity v. This leads to a slowly increasing force (see Equation (3.8), schematically shown in subplot (c)). Both approaches, AFM and SMD lead to force displacement information. In addition to the $F(x)$ curve, SMD provides detailed information about associated atomistic deformation mechanisms. Due to the time scale limitations of MD to several nanoseconds, there is typically a large difference between simulation and experiment with respect to pulling rates. Whereas MD simulations are limited to pulling rates of ≈ 0.1 m/sec, experimental rates are six to eight magnitudes smaller than those. This requires additional consideration before comparing MD results with those from experiments.

3.2 Simulation approach

Here we provide a brief review of interatomic force fields and modeling approaches suitable for simulating the behavior of protein structures. We refer the reader to more extensive review articles for additional information, in particular regarding force field models [87, 88].

3.2.1 CHARMM force field

The basis for most studies presented in this Thesis is the classical force field CHARMM [89, 90], implemented in the MD program NAMD [91]. The CHARMM force field [92] is widely used in the protein and biophysics community, and provides a reasonable description of the behavior of proteins. This force field is based on harmonic and anharmonic terms describing covalent interactions, in addition to long-range contributions describing van der Waals (vdW) interactions, ionic (Coulomb) interactions, as well as

HBs. Since the bonds between atoms are modeled by harmonic springs or its variations, bonds (other than HBs) between atoms can not be broken, and new bonds can not be formed. Also, the charges are fixed and can not change, and the equilibrium angles do not change depending on stretch. The CHARMM force field belongs to a class of models with similar descriptions of the interatomic forces; other models include the DREIDING force field [93], the UFF force field [94], or the AMBER model [87, 95].

In the CHARMM model, the mathematical formulation for the empirical energy function that contains terms for both internal and external interactions has the form:

$$U(r_{ij}) = \sum_{bonds} K_b(b-b_0)^2 + \sum_{UB} K_{UB}(S-S_0)^2 + \sum_{angle} K_\theta(\theta-\theta_0)^2 +$$
$$\sum_{dihedrals} K_\chi(1+\cos(n\chi-\delta)) + \sum_{impropers} K_{imp}(\phi-\phi_0)^2 + \qquad (3.7)$$
$$\sum_{nonbond} e\left[\left(\frac{R_{min(i,j)}}{r_{ij}}\right)^{12} - \left(\frac{R_{min(i,j)}}{r_{ij}}\right)^{6}\right] + \frac{q_i q_j}{\tilde{\varepsilon}_1 r_{ij}}$$

where K_b, K_{UB}, K_θ, K_χ, and K_{imp} are the bond, Urey-Bradley, angle, dihedral angle, and improper dihedral angle force constants, respectively; b, S, θ, χ and ϕ are the bond length, Urey-Bradley 1,3-distance, bond angle, dihedral angle, and improper torsion angle, respectively, with the subscript zero representing the equilibrium values for the individual terms. Figure 3.1 (b) shows a schematic of the individual energy contributions constituting equation (3.7).

The Coulomb and Lennard-Jones 12-6 terms (see Figure 3.1 (a)) contribute to the external or nonbonded interactions; e is the Lennard-Jones well depth and $R_{min(i,j)}$ is the distance at which the Lennard-Jones potential equals 0, q_i is the partial atomic charge, ε_1 is the effective dielectric constant, and r_{ij} is the distance between atoms i and j. The parameters in such force fields are often determined from more accurate, quantum chemical simulation models by using the concept of force field training [96], as illustrated in Figure 3.6.

Force fields for protein structures typically also include simulation models to describe water molecules (explicitly), an essential part of any simulation of protein structures [87, 88].

3.2.2 Input data

We take structures obtained from x-ray diffraction experiments and stored in the Protein Data Bank (PDB) as the starting point for our atomistic simulations. The structure is prepared for the simulations in VMD, and explicit water solvent is added. We perform energy minimization and finite temperature equilibration of all structures simulated before the protein is loaded by applying the SMD technique. The atomistic structures used for the case studies are mentioned explicitly in the individual sections [97].

3.2.3 Steered molecular dynamics

To apply the forces to the molecule that induce deformation, steered molecular dynamics (SMD) has evolved into a useful tool [98]. SMD is based on the concept of adding a harmonic moving restraint to the center of mass of a group of atoms. This leads to the addition of the following potential to the Hamiltonian of the system:

$$U(x_1,x_2,...,t) = \frac{1}{2}k_{SMD}[vt-(\vec{X}(t)-\vec{X_0})\cdot\vec{n}]^2, \qquad (3.8)$$

where $\vec{X}(t)$ is the average position of restrained atoms (x_r) at time t, $\vec{X_0}$ denotes original coordinates and v and $\vec{n}$ denote pulling velocity and pulling direction respectively. The net force applied on the pulled atoms is $F(x_1,x_2,...,t) = k_{SMD}(vt-(\vec{X}(t)-\vec{X_0}))\cdot\vec{n}$. By monitoring the applied force F and the position of the atoms that are pulled over the simulation time, it is possible to obtain force-versus-displacement data that can be used to derive the mechanical properties such as bending stiffness or the Young's modulus (or other mechanical properties). SMD studies are typically carried out with a spring constant $k_{SMD} = 10 \text{kcal/mol/Å}^2$. The SMD method mimics an AFM (Atomic Force Microscopy) nanomechanical loading experiment, as illustrated in Figure 3.2. As shown in this shematic a protein is pulled by a cantilever with a defined speed v along a coordinate x. The deflection of the cantilever (known stiffness k_{AFM}) is a measure of the applied force.

3.2.4 Large-scale parallelized computing

Large-scale MD simulations often require a significant amount of computing resources. Classical MD can be quite efficiently implemented on modern supercomputers using parallelized computing strategies. Such supercomputers are composed of hundreds of individual computers or processors that combined form an entity referred to as supercomputer. Supercomputers exceed the capabilities of ordinary PCs or laptops by several orders of magnitudes.

Whereas computing power was estimated to plateau at the gigaflop level, the broad availability of teraflop computers is now expected by the middle or end of the current decade [12, 99]. Figure 3.3 depicts the development of computational power over several decades, illustrating the emergence of petaflop computers in the next few years.

Based on the concept of concurrent computing, modern parallel computers are made out of hundreds or thousands of small computers working simultaneously on different parts of the same problem. Information between these small computers is shared by communicating, which is achieved by message-passing procedures, enabled via software libraries such as the "Message Passing Interface" (MPI) [100]. Implementations based on spatial domain decomposition allow parallel MD reaching linear scaling, that is the total execution time scales linear with the number of particles $\sim N$, and scales inversely proportional with the number of processors used to solve the numerical problem, $\sim 1/P$ (where P is the number of processors) [100].

With a parallel computer whose number of processors increases with the number of computational cells (the number of atoms, or more general, particles per computational cell does not change), the computational load remains constant. To achieve this, the simulation volume is divided up into computational cells such that in searching for neighbors interacting with a given atom or particle, only the computational cell in which it is located and the next-nearest neighbors have to be considered. This scheme allows to treat huge systems with several billion and more atoms or particles [101]. Alternative approaches, such as computing on Graphical Processing Units (GPUs) provide additional opportunities for extremely high performance. The concept of such approaches is to take advantage of particularly designed processing units to enable ultra-fast operations, tailored to the needs of MD.

Most simulations presented in this Thesis were carried out on a 92 CPU computer cluster "SUNRAY", a Dell machine with Intel Dual Core CPUs.

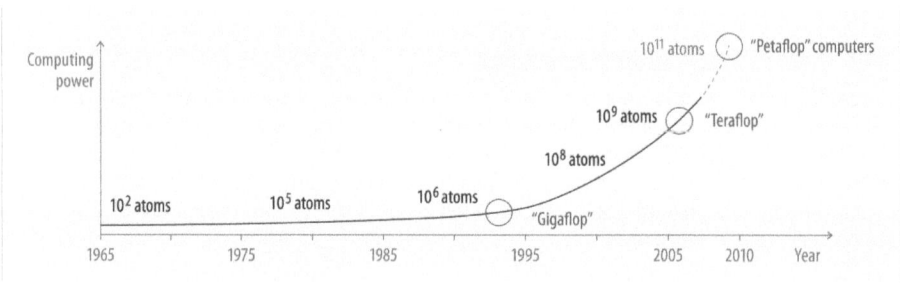

Figure 3.3: Development of computing power over the past decades. The development illustrates the emergence of petaflop computers in the next few years. The plot also summarizes the number of atoms that can be treated with these computing systems; these numbers are developed for simple interatomic potentials with short cutoffs. For CHARMM, the number of atoms is significantly smaller Figure. Adopted from [12].

3.2.5 Data analysis and visualization methods

When analyzing MD simulation data with the focus on mechanics, the calculation of the atomistic stress tensor is of high relevance. The virial stress tensor can be used to calculate the Cauchy stress tensor directly from atomistic data [102, 103]. The atomistic data is averaged over all particles (spatial average), and over several snapshots (temporal average). This approach is used in Section 6.2, in studies of protein networks, for instance. The virial stress is calculated by considering the network volume, including the free volume in the network. For details regarding the calculation of the virial stress tensor we refer the reader to the literature [103]. An alternative method to calculate the stress for filamentous structures is to divide the applied force by the cross-sectional area, $\sigma = F/A$. This approach is analogous to the engineering definition of stress as it is used in the study of tensile tests of single proteins and protein filaments, for instance.

Visualization plays a crucial rule in the analysis of MD simulation results, as the raw data resulting from such numerical simulations represents merely a collection of positions, velocities and accelerations as a function of time. In particular, structural features and patterns of proteins are difficult to analyze. To address this point, many visualization tools exist that are capable of displaying biological protein molecules and clusters therefore. A rather versatile, powerful and widely used visualization tool is the Visual Molecular Dynamics (VMD) program [101]. This software enables one to render complex molecular geometries using particular coloring schemes. It also enables us to highlight important structural features of proteins by using a simple graphical representation, such as alpha-helices, or the protein's backbone. The simple graphical representation is often referred to as cartoon model. These visualizations are often the key to understand complex dynamical processes and mechanisms in analyzing the motion of protein structures and protein domains, and they represent a filter to make useful information visible and accessible for interpretation.

3.2.6 Complementary experimental methods on nanoscale

Recent advances in experimental techniques further facilitate analyses of ultra-small scale material behavior. For instance, techniques such as nanoindentation, optical tweezers, or atomic force microscopy (AFM) can provide valuable insight to analyze the molecular mechanisms in a variety of materials, including metals, ceramics and proteins. The mechanical signature of proteins and other single biomolecules can be obtained by AFM, where the biomolecule is attached to a surface and manipulated by a cantilever that pulls the molecule at constant force rates (see Figure 3.2). A saw-tooth shaped force-displacement profile is commonly observed and linked to sequential unfolding of certain domains in the protein. The worm-like chain model (WLC) [104, 105] is frequently used to describe the entropic elasticity of these domains. We refer the reader to other articles regarding details of these experimental approaches (see, e.g. [35, 106-112]). A selection of experimental techniques is summarized in Figure 7.6, illustrating the overlap with multi-scale simulation methods.

Since recent advances in experimental methods now enable one to probe time- and length-scales that are also directly accessible to large-scale atomistic based simulation, the combination of experiment and simulation might lead to a particularly fruitful interaction. This is higly promising since the kind of information obtained from experiment and simulation might be complementary.

3.3 Hierarchical multi-scale modeling approach

The above mentioned MD simulation techniques provide unique and detailed information about processes and mechanisms at the atomistic level, which cannot be reached by experiment alone. However, even if the simulation codes are improving continuously and computational facilities provide immense calculation capacities, MD techniques still have limitations in their length and time scale (on the order of nm and ns), leaving a gap between the experimental and the simulation scale.

To overcome these limitations, multi-scale modeling techniques are being developed that enable a bottom up description of the material properties, linking nano to macro. In such methods, the model at a coarser scale is trained by a more accurate scale. The concept of integrating various simulation methods by handshaking to bridge across the scales is schematically represented in Figure 3.4. By using this strategy, despite the computational limitations, a link between atomistic resolution and macroscopic time and length scale is possible. We apply this method for mesoscale simulations of protein networks in Section 6.2. In particular for biological materials that feature a high level of hierarchy, this strategy appears to be fruitful. For a detailed review of different coarse graining models, which allow reaching higher time and length scales compared to MD simulations, we refer to the literature [113-115].

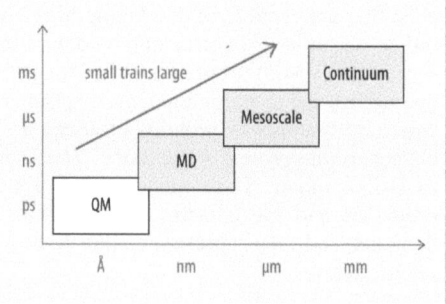

Figure 3.4: Hierarchical multi-scale scheme. Using this first principle approach, where the next level of hierarchy is empirically trained by the previous one, a link between atomistic accuracy and continuum scale is possible. The force-fields applied in MD are derived from quantum mechanical (QM) calculations. MD simulations allow than to study material behavior with atomistic resolution. This method is applied throughout Chapter 5. In Chapter 6.2, we build a direct link from MD simulation to meso-scale simulations, by coarse-graining (see also Figure 6.5 (a)). This allows reaching much higher time and length scales. Results on this scale can be applied to train the next higher scale, the continuum scale. For instance data shown in Figure 6.6 can be used as input parameters for FEM-simulations, allowing making predictions on the macroscale. Relevant experimental methods for each scale are summarized in Figure 7.6.

4 Fracture mechanisms of hierarchical biological materials

In Section 1 we have discussed the overwhelming and intriguing features that exist in biological materials, making them highly interesting as a field of investigation, inspiration and imitation for technological applications.

However, the first step towards engineering structures with similar properties is to derive a better understanding of Nature's material design, and to generate a theoretical link between the hierarchical structures and the observed, resulting properties through the development of a fundamental theory that provides a link between structure and property. This is the particular focus of this Chapter.

4.1 Engineering biological nanomaterials

Mechanical loading of tissues, cells and protein fibers can result in severe changes in the protein structure, inducing HB rupture and protein unfolding at the lowest scale. Large deformation of protein structures can for instance be induced at macroscopic crack-like defects (voids, flaws, soft inclusions) in tissues, where the stresses and thus molecular forces display a singularity (the stress scales as $\sigma \sim 1/\sqrt{r}$, where r is the distance from the crack tip) [12, 116, 117]. At such defects, each protein is exposed to large forces, and at some level the resistance to macroscopic crack growth depends on how much resistance each protein molecule provides at the nanoscale [118]. Therefore, an understanding of the unfolding behavior of proteins is critical on the path towards understanding the fracture mechanics of biological tissues. In this sense, even though many detailed aspects remain poorly understood, protein unfolding and the related HB rupture in structural PMs represent a fundamental crack-tip mechanism, in analogy to dislocations in metals or crack tip mechanisms in brittle materials.

Covalent bonds	Due to overlap of electron orbitals, e.g. found in carbon nanotubes or in the backbone of proteins
Metallic bonds	Found in all metals, e.g. copper, gold, nickel
Electrostatic interactions	Ionic/ coulomb interactions, found n ceramics such as Al_2O_3 or SiO_2 as well as interactions between proteins
Hydrogen bonds	Weak interactions found in polymers or proteins
Van der Waals interactions	Dispersive interactions, e.g. found in wax

Table 4.1: Summary of different kinds of bonds, existing in materials. Table adopted from the book: Buehler, M.J., *Atomistic modeling of materials failure,* Springer, 2008 [12].

Typically, a variety of unfolding processes exist for a given protein structure, each of which has a specific reaction pathway and an associated energy barrier [119], partly related to specific HB breaking mechanisms and rearrangements of the protein structure. Therefore, the key in understanding the unfolding process is first to understand how external forces influence the free energy landscape of the protein, *i.e.* how the relevant energy barriers along the reaction path are shifted by externally applied forces, and second, to capture the dynamics of *soft* HB rupture, which initiates the protein unfolding. HBs are often called soft bonds, as they exhibit three to four orders of magnitude smaller energies and much smaller rupture forces compared to covalent bonds [43, 119, 120]. Table 4.1 summarizes the different kinds of bonds that exist in a variety of materials.

Breaking of interatomic bonds equals to a chemical reaction, which makes an integrated chemomechanical approach essential for a detailed understanding of protein fracture mechanisms and for the development of constitutive mathematical relations. Therefore, for understanding material failure and for engineering biological nanomaterials, the central questions lie at the interface of chemistry, mechanics and thermodynamics: How does a particular protein structure respond to mechanical load? What are the fundamental fracture mechanisms underlying this behavior? How do hierarchies control the fracture and deformation behavior?

Up until now, engineers have developed strength models for bulk materials such as metals, ceramics or polymers, most of which follow continuum approaches [12, 116, 121]. However, in hierarchical nano-patterned materials, the conventional mean-field averaging approach is not applicable, due to an insufficient number of sub-elements [122]. Thus in simple averaging schemes, information may be forfeited that might be crucial for the behavior of several scales above [123], making the explicit consideration of hierarchical features mandatory. Consequently, engineering biological or biologically inspired nanomaterials requires completely new theoretical approaches, *i.e.* by defining the geometry of individual bonds in individual nano-elements, where material strength will be calculated through considering discrete 'ensembles' of bonds. All this shows that existing strength models for materials are not applicable to biological protein structures, and the

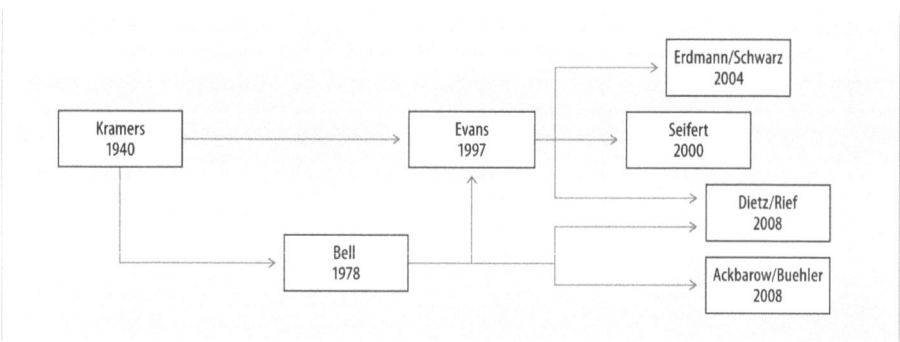

Figure 4.1: Historical evolvement of different theories on bond breaking dynamics. Most of them are derived from a phenomenological theory originally postulated by Bell [20] or Kramer's diffusion model [27]. Extensive work has been undertaken by Evans. He extended the theory by introducing the loading rate as a critical parameter. Further Evans undertook studies on parallel and sequential arrangements of bonds. Seifert and Erdmann/Schwarz studied extensive clusters of parallel bonds. Seifert thereby focused on dynamic loading [39], whereas Erdmann/Schwarz worked with constant forces [40]. Dietz and Rief recently published a paper on networks of bonds, where they describe the elastic behavior of bonds spanned between nodes, where each node represents a C_α-Atom. They assume that bonds are not able to rebind and that the rupture of one bonds leads to the failure of the whole system. The numerical solution of this model allows to describe the anisotropic elastic behavior of proteins, such as the green fluorescent protein [43]. The theory developed by Ackbarow/Buehler is in the focus of this Thesis.

development of such models remains a major scientific frontier.

Only once mathematical relations are developed and validated for PMs, predictive calculations and consequently engineering of protein based materials is possible. In this Chapter we develop a theory, which enables predicting the strength of HBMs in dependence of the strain rate and the structural geometry.

At this point several different theories exist in this field, as reviewed below. However, the main difference between the existing theories and the theory derived in this Thesis is the appearance of the analyzed protein model systems *in vivo:* Earlier theories mainly cover single adhesion bonds or clusters of several parallel or sequential adhesion bonds (e.g. at membranes, focal adhesion in cells, ligand binding to proteins). Here we focus specifically on clusters of HBs in protein filaments, and validate the theory for hierarchical arrangements of parallel HBs and parallel AHs stabilized by HBs rather than long ranging adhesion bonds.

4.2 Previous theoretical work on bond breaking dynamics

Several theories exist that describe competing processes due to mechanically induced instabilities of protein structures (see Figure 4.1 for historical evolvement of different theories). Most of them are derived from a phenomenological theory originally postulated by Bell [20] or Kramer's diffusion model [27]. These models are an extension of the transition state theory for reactions in gases developed by Eyring and others [124] that were inspired by Zhurkov's work on the strength of solids [125]. Here, after reviewing previous work, we extend Bell's approach and show how simulations at various pulling speeds can be used in order to gain information about the free energy landscape of a protein and consequently describe the mechanical behavior.

Systematic studies of protein unfolding at different pulling rates provide valuable insight into the protein's mechanical and thermodynamic stability and behavior. Such studies increase the understanding of the protein's behavior on different time scales, as the pulling speed is not only the key characteristic in defining the time scale of the unfolding process, but determines also the reaction mechanism of the unfolding.

Tremendous contributions in this field have been made by Evans and co-workers [126-133] by developing a theory that describes the binding behavior of proteins in dependence of the loading rate r_f (increase in force over time, here referred as the "loading rate theory"), which up to now was a limitation of the Bell theory. Thereby he related to basic ideas of Kramers [27]. The experimental and simulation data published by Evans let suggest that the energy landscape is governed by multiple transition states, represented by a spline curve consisting of straight lines in the force–log(r_f) graph. However, in the published work no link was made between the calculated parameters E_b and x_b, which describe the free energy landscape and the according mechanisms.

The loading rate r_f is defined macroscopically as the increase in force over time, and can be seen microscopically as the rupturing force divided by the time for bond breaking (reciprocal of the off rate) [126]:

$$r_f = \frac{\Delta f}{\Delta t} = \frac{k_B \cdot T}{x_B} \cdot \omega_0 \cdot \exp\left(-\frac{(E_b - F \cdot x_b)}{k_B \cdot T}\right).$$ (4.1)

The relation between the loading rate and the pulling speed is as follows:

$$r_f = K_0 \cdot v,$$ (4.2)

Here, K_0 is the spring constant of a molecular pulling cantilever.

In contrast to the phenomenological model, Szabo, Hummer, Dudko and co-workers [134-138] follow a slightly different approach in their so-called "microscopic theory". Here, they build their theory on the assumption of only one transition state (single energy barrier). This energy barrier is not only lowered with increasing external force f applied to the molecule (similar to the phenomenological theory), but simultaneously the maximum of the energy barrier is moved along the reaction coordinate towards the equilibrium and eventually vanishes, when the barrier disappears [135] (that is, the parameter x_b changes). This results in a curved instead of a straight line in the force-log(v) space.

In order to derive the correct information about the free energy landscape of the equilibrated system with the microscopic theory, the AFM pulling experiment or the SMD simulation (both non-equilibrium processes) need to be repeated several times, which poses challenges in particular since simulations could be numerically extremely expensive (e.g. a single protein tensile test at low rates could run for 6-8 weeks). These generated data need to be averaged out afterwards, by applying the Jarzynski identity, which postulates that the thermodynamic free energy difference of two states equals to the work along the non-equilibrium trajectory [139, 140].

The microscopic theory postulates that the simple, straightforward phenomenological regime is a very good approximation, but only over a certain magnitude of pulling velocities and lead to overestimations of the off rate at small pulling speeds. The more accurate microscopic approach is in contrast to that more complicated, as an additional parameter (the free energy of activation) is necessary.

Some of the advantages and disadvantages of both approaches were recently summarized in [141-143]. There are other theories related to this topic, which were published recently [144-147].

4.3 Integration of pulling speed in Bell's phenomenological model

4.3.1 Conventional Bell Model

Bell's model is a simple and a broadly applied model, which describes the dissociation rate of reversible bonds [148]. Thirty years ago in 1978, Bell was the first to show the significant role of mechanical forces in biological chemistry by linking the bond off rate (how often a bond breaks per unit time) to externally applied forces, for the case of cell adhesion. His model is an extension of the transition state theory for reactions in gases developed by Eyring [124] in which he has included ideas from the kinetic strength theory of solids postulated by Zhurkov [125].

The off rate χ (the frequency of bond dissociation) is the product of a natural bond vibration frequency $\omega_0 \approx 1 \times 10^{13}$ s^{-1} [20], and the quasi-equilibrium likelihood of reaching the transition state with an energy barrier E_b normalized by the temperature. The energy barrier is reduced by the mechanical energy $F \cdot x_b \cdot \cos(\theta)$ resulting from the externally applied force F (see Figure 4.2 (c)). We note that x_b is the distance between the equilibrated state and the transition state, and θ is the angle between the direction of the reaction pathway of bond breaking (x-direction) and the direction of applied load. The angle can be determined by analyzing the system geometry. The off rate, which equals to the reciprocal of the lifetime of a bond, is thus given by

$$\chi = \frac{1}{\tau} = \omega_0 \cdot \exp\left(-\frac{(E_b - F \cdot x_b \cdot \cos(\theta))}{k_B \cdot T}\right). \tag{4.3}$$

From this equation, it is possible to calculate the force for forced bond breaking within a given time scales of bond dissociation, τ:

$$F(\tau) = \frac{1}{x_b \cdot \cos(\theta)} \left[k_B T \ln\left(\frac{1}{\omega \tau}\right) + E_b \right]. \tag{4.4}$$

A schematic that illustrates the main concepts behind the Bell model is shown in Figure 4.2 (c).

4.3.2 Integration of pulling speed

We begin the derivation with an integration of the pulling speed into the phenomenological Bell model [20]. This is useful since the pulling speed is the key parameter in experiment and MD simulations, allowing linking different time scales.

It is noted that the mathematical symbols used in this Section are summarized in the Appendix.

As shown in equation (4.3), the off-rate, describes how often a bond is broken per unit time (which equals the reciprocal of the lifetime of a bond). The bond breaking speed equals to the distance, which is necessary to overcome the energy barrier divided by the time, leading to the following expression:

$$v = \frac{x_b}{\tau} = x_b \cdot \omega_0 \cdot \exp\left(-\frac{(E_b - F \cdot x_b \cdot \cos(\theta))}{k_B \cdot T}\right). \tag{4.5}$$

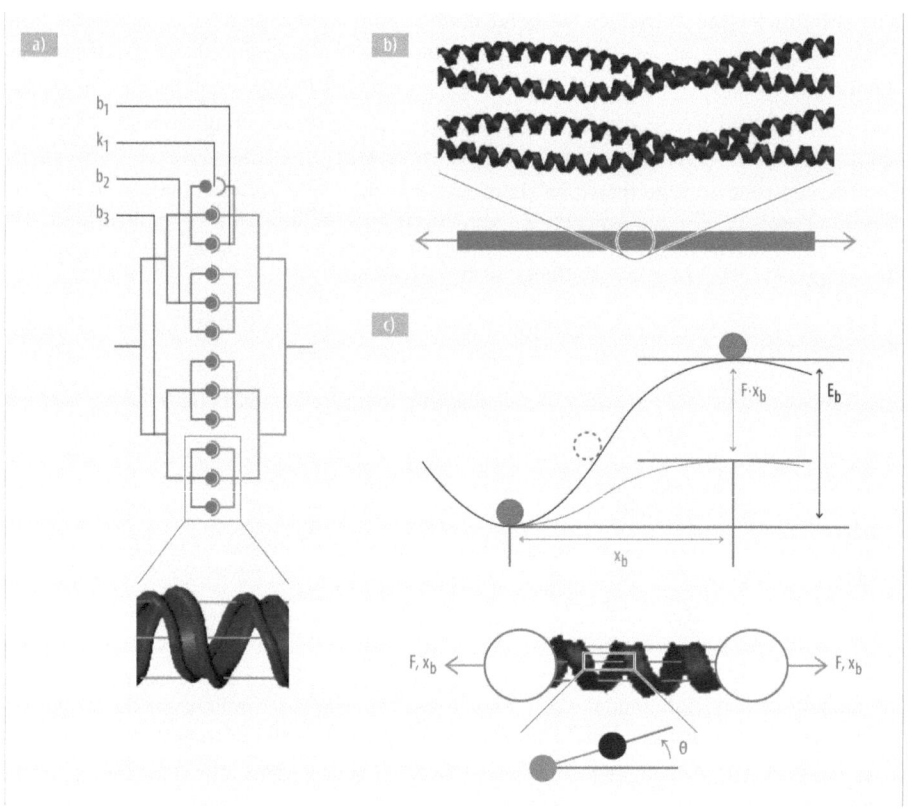

Figure 4.2: Illustration of different hierarchies and their representation in the Hierarchical Bell Model (subplot (a)), as well as representation of the corresponding physical system (subplot (b)). Thereby in subplot (a) the serial arrangement as it exists in (b) is not shown. The inlay in the lower part of subplot (a) shows a single AH structure with ≈ 3 HBs per convolution. The Hierarchical Bell Model reported here enables one to predict the strength of such hierarchical bond arrangements as a function of the deformation speed. Subplot (b) shows the physical system that is represented in the hierarchical model in subplot (a). Subplot (c): Statistical theory to predict the bond rupture mechanics [14]. The graph depicts the energy as a function of deformation along a deformation variable, along a particular pathway that leads to bond rupture. Here F is the applied force, and x_b is the displacement in the direction of the applied force. In the schematic, three HBs (indicated by the red color) break simultaneously. Thus, x_b corresponds to the lateral displacement that is necessary to overcome the bond breaking distance of a HB.

If the protein is pulled faster than the natural bond breaking speed v_0, the externally applied pulling speed equals to the average bond breaking speed. Simultaneously, v_0 defines the speed at which the system is not in equilibrium any more and thus defines the range in which this theory is valid. In general terms: the system must be forced outside the equilibrium in order to measure the resistance of the system.

The link between the theory derived in this Thesis and the loading rate dependent theory developed by Evans [126] is the force F, i.e. the most frequent force at which a system ruptures (defined as the maximal force of a system). Mathematically, both theories can be connected as follows: Following Equation (4.2) we can write the pulling speed as

$$v = r_f / K_0 = \frac{\Delta x}{\Delta t}. \tag{4.6}$$

Combining Equations (4.1) and (4.6) we obtain

$$v = \frac{k_B \cdot T}{K_0 \cdot x_B} \cdot \omega_0 \cdot \exp\left(-\frac{(E_b - F \cdot x_b)}{k_B \cdot T}\right). \tag{4.7}$$

By comparing (4.7) with (4.5), we get

$$x_b = \frac{k_B \cdot T}{K_0 \cdot x_b} \quad (4.8)$$

By moving the spring constant K_0 to the left side of the equation, we receive the expression for the thermal force f_b (equals to the externally applied force $x_b \cdot K_0$), which lowers the energy barrier by one unit of thermal energy $k_B \cdot T$. Already defined for instance by Evans, f_b is an important scaling parameter, as it describes the slope in the force-log-pulling speed curve

$$f_b = x_b \cdot K_0 = \frac{k_B \cdot T}{x_b}. \tag{4.9}$$

Even though the model shown in Equation (4.3) explicitly considers the concept of chemical bonds, it does not distinguish between a single chemical bond and protein architectures that includes several bonds in clusters. For instance, whether a single HB ruptures or if several HBs rupture simultaneously is captured in an effective value of E_b; however, this change in mechanism is not explicitly noted in the theory and thus cannot be predicted (it is simply lumped into an effective value of E_b).

In order to estimate the strength a protein structure without performing any simulations or experiments, here we extend the theory to explicitly consider the structural hierarchies of the protein structure. The only input parameters in addition to the geometry are the energy E_b^0 of a HB and the rupture distance x_b, providing a first principles based description of the protein strength and protein mechanics.

4.4 Hierarchical Bell Model: Considering the hierarchical arrangement

4.4.1 Previous work on multiple bond cluster dynamics

Different suggestions were made for multiple bond descriptions of protein materials mainly distinguishing between parallel vs. serial arrangements of bonds, bonds with and without rebonding as well as soft and hard transducers (transducer-stiffness is relevant for simplifying assumptions regarding the kinematics and for the validation in experiment and simulation). In this study serial as well as parallel arrangements of bonds are considered. However, the validation in MD will be performed only by varying the parallel arrangement of bonds. Further no rebinding is assumed. This assumption is justified, as HBs are in the focus of the theory reported here. Known as short range interactions (in contrast e.g. to adhesion bonds that feature long-range interactions due to electrostatic forces) and being positioned within a distance of 3 Å (equals to the length of an AA residue), apparently HBs tend to break either sequentially (high pulling speeds) or cooperatively (low pulling speeds) rather than randomly, followed by rebonding. Moreover, a stiff force transducer is assumed in our case. The limitations regarding the transducer are reasonable as this is the case for the MD simulation setting, which will be applied for theory validation.

A detailed analysis of the different cases mentioned above was undertaken by Seifert, Erdmann and Schwarz as well as Evans [39, 129, 148-150]. Further, Rief and co-workers recently published a model for a elastic networks of bonds in a protein structure loaded in different directions [43]. For a detailed study of different approaches we refer to the literature. Here only the major scaling relations relevant for the derivations and validation are presented below.

Seifert [39] and Evans [129] suggest that for deformation with a stiff cantilever and irreversible bonds the rupture force scales linearly with the number of bonds b_1 involved.

$$F \sim b_1 \cdot f_b \qquad (4.10)$$

This seems intuitive and was already suggested similarly by Bell. The main assumption behind this equation is that the applied force is shared equally between the existing parallel bonds. If this assumption is not satisfied (e.g. due to particular geometries or loading conditions such as tearing load) other scaling relations appear. For example, Evans analyzed the zipper like tearing failure of bonds, that is, the breaking of bonds in sequence at random times from first to last bond. He calculated with his model that the separation force is less dependent from bond number, due to the logarithmic increase in force with the number of parallel bonds.

$$F \sim \ln(b_1) \cdot f_b \qquad (4.11)$$

A logarithmic but negative scaling relation is observed if l_1 bonds are loaded in sequence. The decrease in force is intuitive due to increasing probability of rupture, as the breaking of one bond is enough for the failure of the whole structure [43, 129].

$$F \sim -\ln(l_1) \cdot f_b \qquad (4.12)$$

Evans also analyzes the extreme case of cooperative rupture of all bonds in a cluster. As all bonds are breaking at once the energy barrier E_b, which needs to be overcome increases by factor b_1 leading to the following scaling relation:

$$F \sim b_1 \cdot f_b \cdot E_b \qquad (4.13)$$

4.4.2 Key assumptions for derivation

As mentioned above, here we assume a non-equilibrium system, a stiff cantilever as well as no rebinding of bonds. Further, following earlier suggestions by Bell, Seifert and Evans we assume that the force of the cantilever F is shared equally between b_1 parallel bonds $f = F/b_1$ of a bond cluster. At this level the bonds can rupture sequentially or cooperatively (due to the geometry, see above). Cooperative rupture of several bonds is in good agreement with the MD studies reported below (see Chapter 5.2.2). Independently, it was also shown that cooperative rupture of several bonds is energetically favorable [151].

Here, in addition to a normal arrangement of parallel bonds in a cluster (hierarchy 1 h_1) we consider for the first time the arrangement of parallel clusters (hierarchy 2 h_2). Thereby we assume that in the arrangement of parallel clusters, which is loaded equally at the beginning, a single cluster will be chosen randomly, which is going to unfold first. This is reasonable, as due to the geometrical arrangement *in vivo* it is improbable that the whole bundle of proteins consisting of several clusters starts to unfold simultaneously. Rather, the unfolding will start in one of the clusters, following a type of zipping mechanism. This might be comparable with the yield mechanism that appears in ductile materials, where deformation appears through sliding of dislocations instead of shearing the whole crystal. This assumption might become not relevant if very stiff elements are present. Simultaneously this assumption is applied as a criterion to distinguish between different clusters. If the kinematics allows a complete rupture and unfolding of a group of bonds independently of an other group of bonds than the two groups can be seen as kinematically independent and thus form two different clusters (see also the beta-helix example at the end of this Chapter).

To the best of our knowledge, a hierarchical arrangement of bond clusters was neither considered in theoretical models nor analyzed systematically with MD simulations until now. Further, a closed form solution, covering both sequential as well as simultaneous rupture of bonds in one equation was not provided. By also considering the length dependence of the structure, this theory allows to build a structure property link for protein materials, an indispensable element towards synthetic design of hierarchical materials, from basic fundamental concepts.

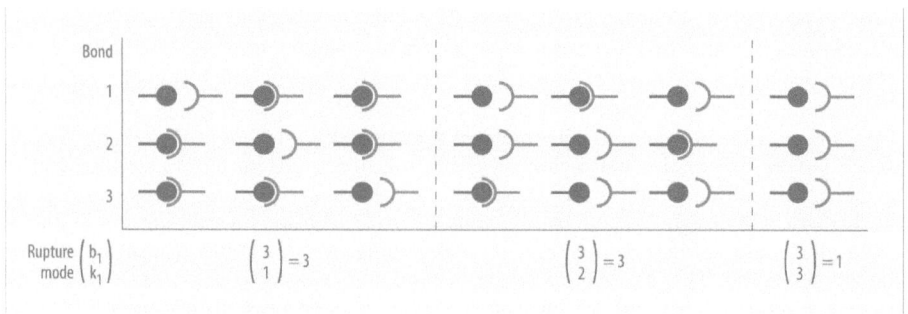

Figure 4.3: Different bond rupture paths for a HB-cluster consisting of three bonds ($b_1 = 3$). The number of bond rupture modes depends on the number of bonds breaking simultaneously (k_1) and is calculated for the example of AHs. We assume that after the initiation of rupture the non-ruptured bonds will rupture very short time later, leading to the unfolding of the cluster (convolution).

4.4.3 Derivation for a simple bond cluster

In this spirit, we thus begin at the lowest hierarchy, which is represented by individual HBs with an E_b^0 and x_b (we refer to it as h_0, as it is the "basic hierarchy"). The next higher hierarchy consists of parallel HBs (hierarchy level h_1). Here we assume that b_1 bonds in a structure are in parallel and k_1 bonds out of these b_1 bonds break simultaneously (see Figure 4.2 (a)). Consequently, b_1 over k_1 possible combinations exist for the rupture mechanism at h_1. This is comparable with different rupture paths for system failure. Figure 4.3 shows the rupture paths in a simple example with $b_1 = 3$. Thereby we assume that all bonds are equal and no distinction between them is possible. The occurrence of one of the shown combinations will initiate failure of the cluster. Consequently, the probability that one of these combinations constitutes a particular rupture event is one divided by b_1 over k_1. In other words, the more equally loaded bonds are in parallel the smaller becomes the probability of system failure, making the structure overall stronger. In contrast to that, for a sequential arrangement of bonds the probability of rupture will increase with the number of bonds, resulting in a factor l_1 (more precisely l_1 over 1) instead of a divisor.

Accordingly, if k_1 bonds break simultaneously, we have to consider that the energy barrier E_b^0, representing the energy barrier of a single HB, increases by a factor k_1 leading to the new energy barrier, which need to be overcome by the system force F. This leads to the following expression for the off rate:

$$\chi_{h1} = \omega_0 \cdot l_1 \cdot \binom{b_1}{k_1}^{-1} \cdot \exp\left(-\frac{k_1 \cdot E_B^0 - F \cdot x_b \cdot \cos(\theta)}{k_B \cdot T}\right). \tag{4.14}$$

We note that already Erdmann and Schwarz applied binomial coefficients when describing the probabilities of states during the stochastic process of bond rupture, in which k_1 out of b_1 bonds are bonded [40]. However, in contrast to this theory, Erdmann and Schwarz assume static load and bond rebinding.

For example, the extreme case of simultaneous bond ruptures of all bonds in a cluster ($k_1 = b_1$) will result in the following expression for the off rate:

$$\chi_{h1} = \omega_0 \cdot l_1 \cdot \binom{b_1}{b_1}^{-1} \cdot \exp\left(-\frac{b_1 \cdot E_B^0 - F \cdot x_b \cdot \cos(\theta)}{k_B \cdot T}\right). \tag{4.15}$$

We rewrite equation (4.14) so that the binomial coefficient appears in the exponential, which enables us to compare Equation (4.16) with Equation (4.3),

$$\chi_{h1} = \omega_0 \cdot \exp\left(-\frac{\left[k_1 \cdot E_b^0 + k_B \cdot T \cdot \left(\ln\binom{b_1}{k_1} - \ln(l_1)\right) - F \cdot x_b \cdot \cos(\theta)\right]}{k_B \cdot T}\right). \tag{4.16}$$

Consequently, the effective energy barrier E_b in Equation (4.3) can be split up in the following way:

$$E_b = k_1 \cdot E_b^0 + k_B \cdot T \cdot \left(\ln\binom{b_1}{k_1} - \ln(l_1)\right), \tag{4.17}$$

where E_b^0 is the energy of a single bond and the term $k_B \cdot T \cdot \left(\ln\binom{b_1}{k_1} - \ln(l_1)\right)$ is the "additional increase" of the energy barrier due to the hierarchical structure, allowing us to predict the effective height of the overall energy barrier. The off rate χ_{h1} can be linked to the pulling speed in the same way as the previously used off rate χ in Equation (4.5). Developing the equation for the force F and including the thermal force $f_b = \frac{k_B \cdot T}{x_b \cdot \cos(\theta)}$ leads to the following expression for the unfolding force, or in more general terms, the "system breaking" force:

$$F^{h1} = f_b \cdot \left(\ln\left(\frac{v}{x_b \cdot \omega_0}\right) + \ln\binom{b_1}{k_1} - \ln(l_1) + \frac{k_1 \cdot E_b^0}{k_B \cdot T}\right) = F_v + F_{h1} + F_{h0}, \tag{4.18}$$

where the F_v, F_{h1} and F_{h0} are the contributions to the force as a consequence of the pulling speed v, the first hierarchy (number of parallel and sequential bonds b_1 and l_1), and the basic hierarchy (strength of bonds, E_b^0 and x_b). This expression quantifies how the hierarchical design influences the rupture strength, by providing explicit expressions for the contributions at each hierarchical scale.

We now compare the predictions of our model with extreme cases considered in earlier models. Analyzing the extreme cases of sequential bond rupture ($k_1 = 1$, corresponding to a tear zipping mode) we obtain the following scaling relation as a function of the number of bonds:

$$F^{h1}_{sequential} \sim \ln(b_1) \cdot f_b \tag{4.19}$$

This result agrees with that of Evans, as we have a weak logarithmic relation due to the zipping, which takes place after initiation of bond rupture (starting with one bond).

For the other extreme case of cooperative bond rupture ($k_1 = b_1$, assuming that all bonds in a cluster break concertedly), we obtain the following scaling relation:

$$F_{cooperative}^{h1} \sim b_1 \cdot f_b \cdot E_b^0 / (k_b \cdot T) \tag{4.20}$$

We note that in case of cooperative bond rupture and equal load distribution the system force increases linearly with f_b and E_b. This is the strongest increase in force with each additional bond compared to the other extreme cases. The factor $E_b^0/(k_b \cdot T)$ is an additional multiplier of about 10 (for a reasonable value of $E_b^0 = 6$ kcal/mol). Further, the scaling relation for the sequential arrangement of bonds also agrees with previous work. This case is optimal from the mechanical perspective; however, there may be thermodynamic limitations to achieve the rupture of a very large number of bonds [151].

4.4.4 Derivation for hierarchically arranged bond clusters

The expression derived above for two hierarchies can be extended in a similar way to a second or higher level of hierarchies ($h_2, h_3, ..., h_n$), which enables one to predict the unfolding rate for example of a tertiary structure consisting of $i=2, 3, ... n$ filaments., of which k_i elements of the previous scale h_{i-1}, unfold simultaneously. Similar as for the two hierarchy system, for a system with three hierarchies, b_2 over k_2 possibilities exist that the unfolding appears in k_2 of the b_2 elements. However, as mentioned in the assumptions at the beginning of this Section, k_i is set to one for the hierarchical level $i > 1$. This assumption might differ if stiff elements are used. Additionally, since the unfolding can begin in k_2 out of b_2 helices (as all of them have initially the same strength), the probability on the next smaller hierarchy h_1 is decreased by the exponential b_2 over k_2 (multiplying b_2 over k_2 times the probability of rupture on h_1), resulting in $\binom{b_1}{k_1}^{-\binom{b_2}{k_2}}$. As we assume that only one cluster unfolds at one time, the exponential Arrhenius-term does not change compared to the previous derivation with only one hierarchy. Finally, we arrive at the following expression for a h_2 system (e.g. CCs):

$$\chi_{h2} = \omega_0 \cdot \binom{b_2}{k_2}^{-1} \cdot l_1 \cdot \binom{b_1}{k_1}^{-\binom{b_2}{k_2}} \cdot \exp\left(-\frac{k_1 \cdot E_B^0 - F \cdot x_b \cdot \cos(\theta)}{k_B \cdot T}\right). \tag{4.21}$$

We note that for simplicity the term l_2 was not mentioned (e.g. resulting from a serial arrangement of CCs, where each element CC consists of b_2 parallel bond clusters), but can be integrated in a very simple way. This is because we mainly focus on parallel rather than sequential arrangements of clusters here.

This equation can be generalized for a system consisting of n hierarchies in the following way:

$$\chi_{hn} = \omega_0 \cdot \binom{b_n}{k_n}^{-1} \cdot \prod_{i=2}^{n} \binom{b_{i-1}}{k_{i-1}}^{-\binom{b_i}{k_i}} \cdot l_1 \cdot \exp\left(-\frac{k_1 \cdot E_B^0 - F \cdot x_b \cdot \cos(\theta)}{k_B \cdot T}\right). \tag{4.22}$$

According to equation (4.17), the system's energy barrier E_b from Equation (4.3) can be written in the following way for an n-level system:

$$E_b^{hn} = k_1 \cdot E_b^0 + k_B \cdot T \cdot \left[\sum_{i=2}^{n} \binom{b_i}{k_i} \cdot \ln\binom{b_{i-1}}{k_{i-1}} - \ln(l_1) + \ln\binom{b_n}{k_n} \right] \qquad (4.23)$$

We note that the additional coefficient in front of the exponential in Equation (4.22) can be summarized to the number of "effective" bonds of the hierarchical system, which increase the lifetime of a system. Consequently, the second term in Equation (4.23) could be replaced by $k^* \cdot E_b^0$, where k^* would represented the additional "effective" bonds.

With these expressions we are now able to predict the force of an n-level system, where the force contribution at each hierarchical scale is considered explicitly:

$$F^{hn} = f_b \cdot \left(\ln\left(\frac{v}{x_b \cdot \omega_0}\right) + \ln\binom{b_n}{k_n} + \sum_{i=2}^{n} \binom{b_i}{k_i} \ln\binom{b_{i-1}}{k_{i-1}} - \ln(l_1) + \frac{k_1 \cdot E_b^0}{k_B \cdot T} \right) = F_v + \sum_{i=0}^{n} F_{hi} \qquad (4.24)$$

We observe from this equation for the case of $n = 2$ that the force scales logarithmically with the number of parallel clusters and under the assumption that in each cluster all bonds break at once. This is in good agreement with the zipping behavior, regarding the level of b_2, which we would also expect. However, the relation is dominated by b_1 parallel bonds due to the linear dependence as well as the additional factor E_b^0

$$F^{h2}_{cooperative} \sim (\ln(b_2) + b_1 \cdot E_b^0 / k_B \cdot T) \cdot f_b . \qquad (4.25)$$

If we assume that in each cluster the bonds break sequentially, we get the following relation.

$$F^{h2}_{sequential} \sim (\ln(b_2) + b_2 \cdot \ln(b_1)) \cdot f_b . \qquad (4.26)$$

In this case the behavior is mainly linear in dependence of b_2. This is an effect of the hierarchical arrangement present in this system.

These equations now enable us to estimate the unfolding force at any pulling speed once the structural hierarchical geometry is known. Due to the generic approach, this equation is valid for any protein structure that consists of several parallel bonds on the basic scale up to systems with several sub-elements on higher hierarchical scales, such as assemblies of AHs, beta-sheets or beta-helices. However, here we focus on the detailed analysis of simpler systems of AH-based assemblies.

In addition to allow for an analytical, continuum-type expression to describe the unfolding dynamics, the Hierarchical Bell Model enables us to bridge time-scales, because we describe the rupture force as a function of the pulling speed. The pulling speed is a controlled parameter in simulation and experiment, which defines the timescale of forced rupture of a protein constituent. Understanding the effect of the pulling velocity on the unfolding behavior is of great significance since protein structures display a strong dependence of the unfolding mechanics on the pulling speed. This may control some of their biological functions, for instance in mechanotransduction or in light of the IF's security belt role. Further, this insight will eventually enable to develop continuum-type descriptions of the viscoelastic properties of proteins. As pointed out in [24], understanding the rate dependent properties is also vital to predict dissipative properties

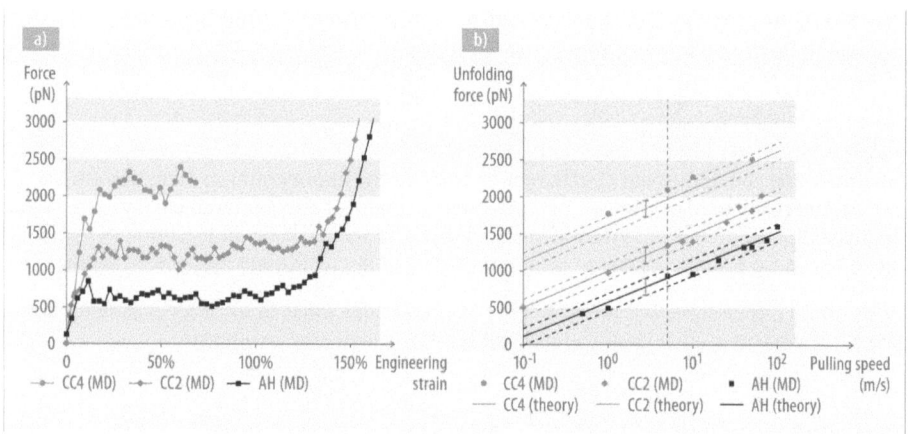

Figure 4.4: Force-extension curves (subplot (a)) and unfolding force (subplot (b)) as a function of pulling speed, carried out for validation of the Hierarchical Bell Model. These curves correspond to the geometries shown in Figure 6.1. The beginning of the plateau regime (here at a pulling speed of 5 m/s) defines the unfolding force, the quantity plotted in subplot (b). Subplot (b) further shows an error bar with a 5% variation in the E_b^0 value.

of proteins, protein networks and cells under cyclic loading as it appears e.g. in the heart muscle. In the next Section we validate and discuss the limitations of this theory.

4.5 Validation and limitations of Hierarchical Bell Model

4.5.1 Validation by direct atomistic simulations

We validate our theory for the AH protein motif. We pick three levels of hierarchies: (i) single AH that consists of a helical polypeptide with 3.6 HBs per convolution (hierarchy 1), (ii) CC proteins (hierarchy 2) that consist of two helically arranged AHs (CC2), and (iii) four-stranded CCs that consist of four AHs arranged in a helical fashion (CC4).

The three structures are shown in Figure 6.1. In order to avoid possible length effects, all proteins have the same length of 7 nm.

For each structure, we carry out a pulling experiment using MD, for varying pulling velocities. Results of these calculations are shown in Figure 4.4, including a quantitative comparison of the MD results with the predictions by the Hierarchical Bell Model. For the predictions by the Hierarchical Bell Model, we first estimate the energy barrier (E_b and x_b) for the single AH. It is found that the controlling deformation mechanism is the rupture of a single HB. This information is then used to directly predict the behavior of more complex structures, without any additional parameter fitting. These results clearly corroborate the concepts proposed in the theory. In addition to the precise value for E_b^0=5.83 kcal/mol, we add an error bar with +/-5% deviation, i.e. for 5.53 < E_b^0 < 6.12 kcal/mol. The E_b^0 value calculated here is in good agreement with previous theoretical [93] (the theoretical estimate for HB energy is between 5 and 10 kcal/mol) and experimental results. While the prediction for the CC2 structure is in excellent agreement with simulation, the results for the CC4 are slightly higher. One explanation could be that this is a structure taken from a completely different protein family that could feature a slightly higher E_b^0 (E_b^0 would be 7.38 kcal/mol to match the observed MD simulation results; still a reasonable value for the HB energy). Another explanation might be the geometry. In the CC4 structure, HBs situated at the inner site are less exposed to water, which might lead to increased values for the effective E_b, making the structure overall stronger than our prediction.

Further validation of this concept is done in Chapter 5 where we apply this theory on different case studies over several length scales.

We note that a systematic validation of this theory with MD is very tedious and costly (if not impossible with currently available computational resources), not only due to missing protein structures (e.g. with different number of AHs per CC), but much more as it requires a high number of simulation runs at varying pulling rates. We therefore suggest a detailed validation with a mesoscale model, where different architectures can be realized much easier. This is work that could be carried out in the future.

4.5.2 Model limitations

Nevertheless, this model - as every model - is a simplistic approximation of reality, neglecting many aspects that appear in reality but focusing on the important physical features of interest. Therefore the model developed here is only valid under certain conditions and thus has limitations, which are discussed below.

Boundary conditions and geometric limitations

This model is only applicable when the system is loaded with tensile stress. Compression, shearing or bending load are not analyzed in detail, and for predictions under these loading conditions additional parameters or modifications may be necessary. Further, we assume that the load is applied at the ends of the molecule and the resulting deformation is distributed initially uniformly over the existing elements. The model allows only describing parallel arrangements of bonds, which are aligned along the protein axis (which is assumed to be the direction of load) or which have a uniform tipping angle relative to the protein axis, as e.g. the case for AHs. Geometries where the bonds have different directions or even different combinations of distinct protein elements (e.g. AHs and BS) cannot be covered by the current formulation of the theory. First efforts in theoretically and numerically governing multidirectional bonds were undertaken recently by Dietz and Rief [43], where they have modeled elastic bond networks.

Biological relevance of modeled structures

In this model only the secondary structure and higher hierarchical levels, and here especially the geometrical arrangement of HBs, are taken from the protein structure, as these parameters primarily define the mechanical properties. It will be shown in a case study in Section 5.1 that changes in the primary structure do not influence the mechanical behavior of the protein as long as the mutation does not destroy the (*in vivo* existing) secondary structure. Therefore, information from the AA sequence does not directly enter the model but is rather captured in the value of the energy barrier E_b^0 for HB breaking (which does depend on the AA sequence). Even if we assume for simplicity that the

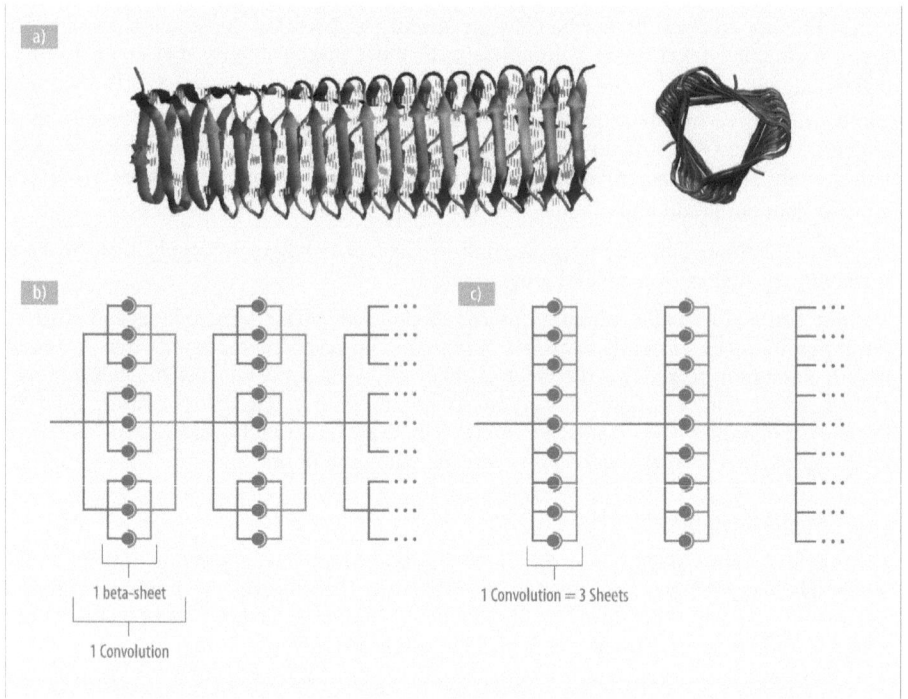

Figure 4.5: Plot shows the applicability of our theory to a different secondary structure, in this case a beta-helix, here shown for a needle of a T4 Bacteriophage (subplot (a) depicts the ribbon structure). Each convolution consists of three beta-sheets arranged in an equilateral triangle, where each beta-sheet consists of three HBs. Subplots (b) and (c) depict two different model structures, which could be derived from the protein structure. Subplot (b) depicts a structure consisting of two hierarchies per convolution, whereas (c) is a simple one-hierarchical system consisting of nine parallel bonds. Our theory predicts that the high of the effective energy barrier E_b of system (b) is 28% higher than the one of system (c). However, it mainly depends on the kinematics restrictions, *i.e.*, if single beta-sheets can unfold independently or not, which hierarchical structure can be used as a model approximation for theoretical predictions. Detailed studies on this protein structure will follow in the future.

geometry as well as the packaging of the analyzed structure are known and can be represented by this model, it is not negligible that finally individual AA sequences can be crucial in determining the geometry and protein packaging (e.g. forming n-stranded CCs) by creating bonds or salt bridges between individual elements, clusters and strands. Thus our model can not make predictions directly from the AA sequence to structural properties.

Another limitation is the fact that only symmetric and periodic structures are represented, *i.e.* systems, where each element consists of several equal sub-elements. However, under *in vivo* conditions this is only the case for a few hierarchical levels and structures, e.g. for AHs, CCs and filaments consisting of several CCs in IFs. Higher-scale hierarchical levels, e.g. networks of filaments, cannot be covered by this theory and other theoretical models are necessary. Nevertheless this theory allows studying hypothetical structures with n hierarchies (possible for $n \to \infty$), which could be of relevance for studies related to synthesis of new hierarchical materials or to understand the behavior of fundamental building blocks of more complex structures (as illustrated in Section 6.2, where studies of the behavior of lamin AH networks are discussed). Another limitation, which arises even if the structure and the number of parallel elements are known, is the missing information about the number of bonds or elements that break simultaneously. This information can only be derived from MD simulations or experiments or other theoretical models (such as Keten & Buehler's thermodynamic HB fracture model [151, 152]). However, once known, different rupture behaviors and rupture paths can be covered by this model.

Other limitations of the model

It was shown by theoretical analysis, MD simulation as well as experiments [141, 143] that the simple, straightforward phenomenological Bell model is a very good approximation, but only over a certain magnitude of pulling velocities, overestimating the off rate at vanishing pulling speeds. As will be shown later, the model is only applicable for pulling velocities, which are higher than the natural bond breaking speed v_0, *i.e.* in the regime of forced unfolding, resulting in negative forces for lower pulling speeds than v_0 (for very low deformation speeds or large time-scales, protein unfolding is no longer controlled by rupture of HBs as a statistical process but rather by the thermodynamic free energy landscape and conformational changes of the polypeptide backbone). Additionally, in this model bond rebinding is not taken into account. However, in this considered deformation speed range, where bond rupture is followed by protein unfolding, this assumption is well justified. Further, this model gives estimates about the energy maximum and its distance to equilibrium, without providing information about the shape of the energy well, which might be crucial for certain experiments or analysis. In this case more complex models with several additional parameters need to be considered [126-133, 144-147], which e.g. take into account that the transition state distance x_b is influenced by the applied force.

In summary, these limitations reduce the applicability of the developed model. However, at the same time it provides a simple, straightforward approach to interpret MD and experimental results in order to derive a fundamental understanding of the protein behavior, the overall effective energy barriers and others, thereby being a middle ground between the crude, purely phenomenological model and highly complex analytically theories that do not provide a direct link between fundamental physical parameters and protein rupture mechanisms. Further the here presented model enables for the first time a structure-property link, which is of great significance when developing and analyzing new structures.

Applicability to other protein structures

Here, we show how our theory can be applied in order to calculate the mechanical properties of a completely different secondary protein structure. We apply it to a triple-beta helix of a T4 bacteriophage, as shown in Figure 4.5 (a). In a first step, an abstraction of the system as a model system in the representation used in our theory is necessary. As shown in Figure 4.5 (b) and (c) different hierarchical arrangements can be considered. In the first possibility (Figure 4.5 (b)), the BS on each side of the triangular convolution are kinematically independent, *i.e.* each BS can unfold independently. If the sheets are kinematically dependent, *i.e.* the rupture of one BS leads automatically to the destruction of the other two BS in this convolution than the schematic with only one hierarchy (Figure 4.5 (c)) is more accurate. Applying Equation (4.23), an $E_b^0 = 5\,\text{kcal/mol}$ per HBs and assuming that the bonds rupture one by one (as e.g. expected in simulations and observed for other structures at high pulling rates), our model predicts that the introduction of the hierarchical level (Subplot (b)) increases the effective energy barrier E_b by 28% compared to the arrangement with only one hierarchy (Subplot (c)). This would consequently result in higher breaking forces, which can be calculated in dependence of the pulling speed by applying Equation (4.24).

This example illustrates the advantages of a theory that provides a direct structure-property link and thus enables us to carry forward case studies that experiment with different structural representations. However, detailed validation for a specific structure and other structures are crucial to identify the molecular details of unfolding processes and kinematic constraints and elements not captured by this model.

5 Mechanics of hierarchical alpha-helical structures, from nano to macro

For four case studies with different protein structures with different hierarchical levels (primary, secondary, tertiary and quaternary structure), we demonstrate how MD simulation paired with the Hierarchical Bell Model described in the previous Section can serve as a powerful tool in increasing our understanding on the nano-mechanical fracture behavior as well as the biological function of abundant protein structures. Figure 5.1 provides an overview on the main research questions addressed at different length scales.

It will be shown in these studies that each AA has multiple roles ranging from the creation of individual HBs up to higher level interactions.

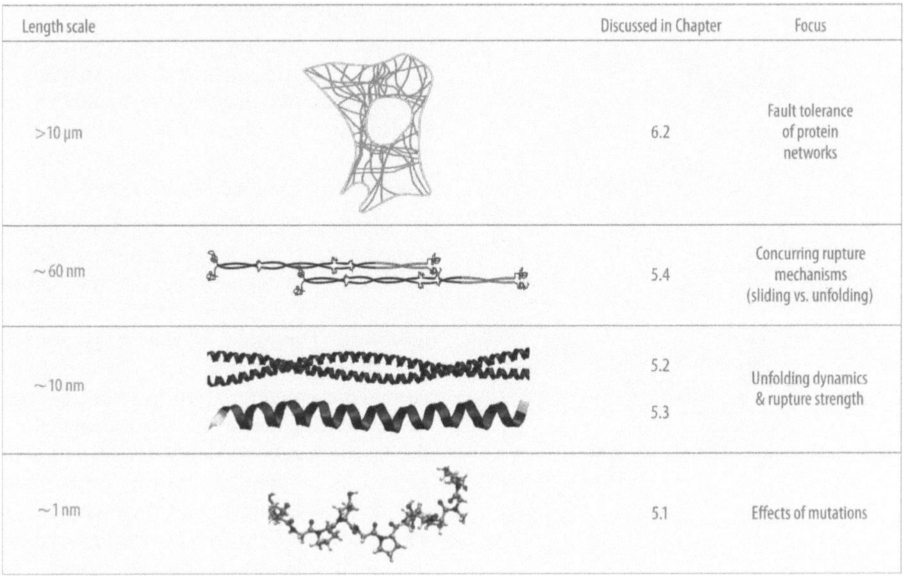

Length scale		Discussed in Chapter	Focus
>10 μm		6.2	Fault tolerance of protein networks
~60 nm		5.4	Concurring rupture mechanisms (sliding vs. unfolding)
~10 nm		5.2, 5.3	Unfolding dynamics & rupture strength
~1 nm		5.1	Effects of mutations

Figure 5.1 Overview over the different length scales studies in this thesis as well as the main focus of the undertaken research. Beginning with studies on the primary structure we will go up the scale up to protein networks as they appear in cells.

5.1 Studies of primary protein structures: Point mutations

In this case study we will analyze the effects of mutations on the mechanical properties of AH structures, mainly with the focus on silencing and activation, *i.e.* how point mutations (do not) cause changes on different hierarchical levels.

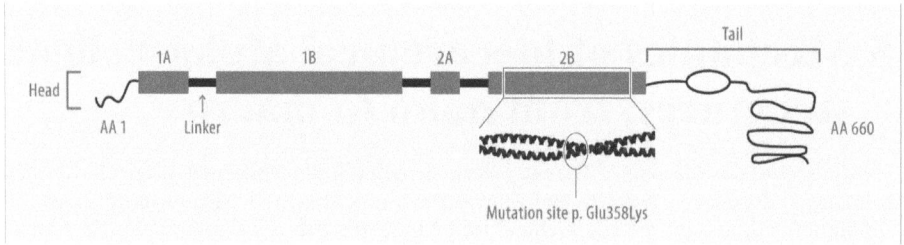

Figure 5.2: Schematic of the Lamin and the CC structure, adapted from [21]. Human lamin is 660 amino acids long. It consists of a head, a tail, and four rod domains. The four rod domains are connected with linkers, whose structure is largely unknown as of today. The study reported in this Chapter is focused on amino acid 313 to 386, the coiled coil structure. The blow-up of the 2B segment shows the CC geometry; a tensile load is applied at the ends of this domain to probe the mechanical properties. The specific mutation considered here is p. Glu358Lys (location indicated in the plot). The dimer building block shown here assembles into a complex hierarchical arrangement of building blocks (not shown here) and is mechanically loaded along the protein axis *in vivo*.

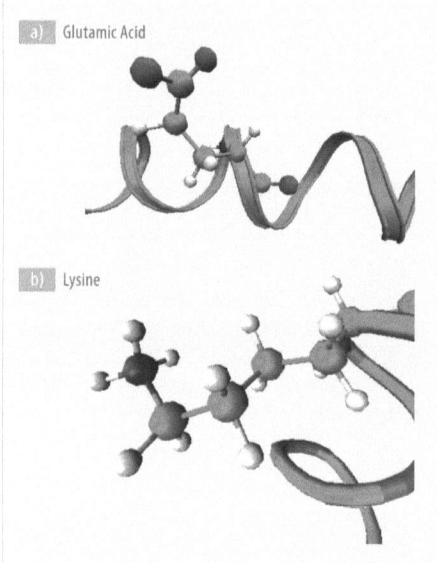

Figure 5.3: Here we consider the mutation p.Glu358Lys, a change from glutamic acid to lysine. This plot shows the structure of the wild type protein (subplot (a)) and the mutated protein (subplot (b)). The plots show the relaxed structure of the area of the protein that contains these side chains, solvated in water (water molecules not shown for clarity). Color code: Blue-nitrogen, red-oxygen, gray-carbon, white-hydrogen. Subplot (a): Glutamic acid has an oxygen atom at the end of the side chain, which is acidic and features a negative charge in water. The side chain has a pKa of 4.07. Subplot (b): Lysine has a nitrogen atom at the end of the side chain, which is basic and features a positive charge in water. The side chain has a pKa of 10.53.

5.1.1 Protein structure

This study is focused on the specific case of muscle dystrophies and the analysis of the mechanical integrity of lamin at the molecular, the dimer level (Figure 5.2). Details regarding the biological function were already provided in Section 2.5.

It is rather challenging for experimental scientists to create single point mutation in a protein and then probe the mechanical properties of a single protein domain or molecule. Therefore, in our study we take an alternative approach by utilizing computational simulation to create a single missense mutation in the 2B rod domain of the lamin A protein (PDB ID 1X8Y). This allows us to simulate tensile experiments on both wild type and mutated lamin. These tensile deformation studies provide us with detailed insight about the nanomechanical effects of particular point mutation on the mechanical properties of individual dimers.

Earlier studies have shown that lamin A/C appears to be more strongly correlated to the nuclear mechanics than lamin B [153]. The screening of patients with Emery-Dreifuss Muscular Dystrophy has clearly shown that the mutation p.Glu358Lys is commonly related to this disease [78, 154].

We therefore chose the mutation at AA 358 of lamin A/C as the focal point of the present study. This mutation, named p.Glu358Lys is a change from a glutamic acid residue to a lysine residue, as shown in Figure 5.3.

The pKa of the glutamic acid side chain is 4.07, and the pKa of lysine side chain is 10.53. Thus the mutation introduces a major change in the charge of the side chain, modifying the positive partial charge in the side chain to a negative partial charge. Changes in the charge of side chains could potentially interrupt salt bridge and affect the structure of protein assemblies and thus influence their mechanical properties on the intermolecular level.

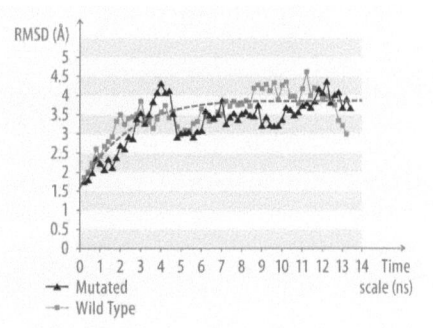

Figure 5.4: RSMD (root mean square deviation) of wild type and mutated lamin dimer through equilibration at 300 K, embedded in explicit water solvent (during equilibration, no load is applied to the molecule). Both structures have been equilibrated for 13.5 ns at 300 K. The RSMD is calculated every 0.2 ns. This plot shows that both structures have converged to their equilibrium structure within the MD time interval. The final structures after equilibration intervals of 13.5 ns are used as input structures for the pulling simulations.

The simulations and analyses reported in this Section are based on the consideration of two model systems, model A, the wild-type lamin structure, and model B, the mutated lamin structure. We generate the mutation, p.Glu358Lys on the wild type lamin by using the mutation function of VMD [101] (see Figure 5.3). Then, we first minimize the energy and then equilibrate both structures for 13,5 ns. The duration of the equilibration at 300 K is dictated by the requirement that the structures converge to a stable configuration. Thus we monitor the root mean square distance (RMSD) of both wild type and mutated lamin during the equilibration phase. In both cases, we observe that the value of RMSD converge to a constant value, as shown in Figure 5.4. This indicates that the structures have approached a stable molecular configuration. This careful equilibration analysis is critical to ensure that the starting configuration is in equilibrium, in particular since we have introduced the mutation into the wild type protein structure.

5.1.2 Results of molecular modeling

We perform tensile simulations by holding one end of the lamin dimer fixed, and pulling the other end of the lamin dimer with a constant speed. We perform simulations under the following pulling speeds: 1 m/s, 5 m/s, 10 m/s and 20 m/s.

Force-displacement plots of tensile experiments are shown in Figure 5.5, for four different pulling velocities. For all pulling velocities, we find three regimes: The first regime is an elastic regime (behaves close to a linear relationship for small deformation) in which the force increases linearly with applied strain. This regime is followed by a plateau regime, leading to the final regime in which the force increases rapidly with increasing displacement. The AH structure is lost during deformation in the plateau regime, beginning at the angular point (AP) between the first and the second regime.

Most notably, by comparing the tensile simulation results of the wild type structure and the mutated lamin structure, we observe no significant difference neither in the force-

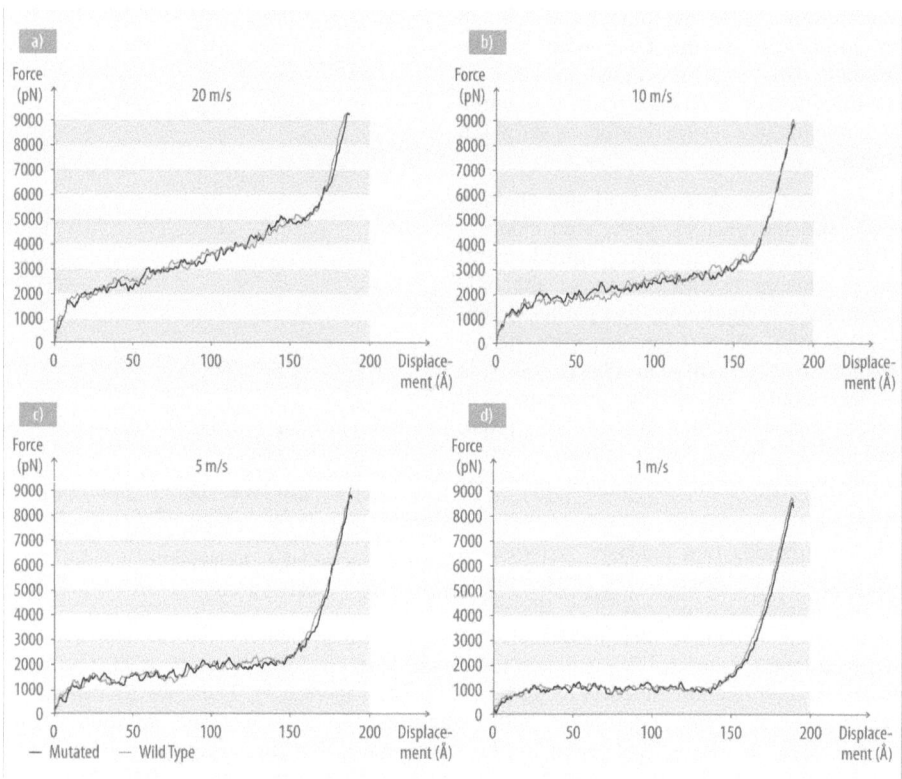

Figure 5.5: Simulated tensile experiments of the lamin dimer domain, at various pulling velocities (subplots (a) 20 m/s, (b) 10 m/s, (c) 5 m/s, and (d) 1 m/s), for both model structures. Both wild type and mutated lamins have been equilibrated for 13 ns at 300 K (see also Figure 5.4 for the corresponding RMSD data). The tensile experiments were also performed under constant temperature of 300 K. To enable pulling, we fix one end of the protein and pull on the other end.

strain data nor the unfolding dynamics (not shown) between wild type and mutated lamin. As clearly shown in Figure 5.5, this behavior is observed consistently for the four different pulling velocities, suggesting that the pulling speed does not influence this behavior and also provides a statistical validation of the observed effects.

Figure 5.6 shows a zoom into the small-deformation regime in the case of $v = 1$ m/s. These results visualize that the behavior in both cases is very similar also for the small-deformation regime, which is the most significant for strains that appear *in vivo*.

Further, the unfolding mechanisms – continuous rupture of HBs – are identical in both the wild type and the mutated protein structure. This has been confirmed by analyzing the rupture mechanisms in VMD (results not shown).

5.1.3 Conclusion in light of materials science and biological function

Point mutations cause defects on higher hierarchical scales

In this Section, we have explored the effect of a single point mutation in lamin dimers. The most important result of this study is that a single missense mutation does not alter the mechanical properties of the lamin rod domain on dimer level. Our results suggest that the mutation most likely affect larger-scale hierarchical features and properties in the lamina network, such as the dimer or filament assembly or even gene regulation processes. Our results provide rigorous atomistic-scale support for earlier claims that have suggested that laminopathies are closely related to higher hierarchical assembly levels of the lamin [34]. Similarly, it was shown recently for gamma D-crystallin that the effects of point mutations are visible at higher hierarchical scales [155].

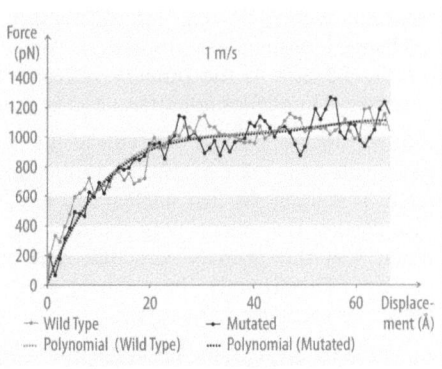

Figure 5.6: Zoom into the small deformation regime of a force-extension curve for a pulling velocity of 1 m/s. The solid lines are 4th order polynomial fits to the two data sets. This analysis provides additional evidence that the behavior of both structures is very similar.

We note that our study has several limitations: For instance, we have only considered a small part of the overall dimer structure. Further, our model did not enable us to study the behavior of several dimers into larger hierarchical assemblies (this could be addressed in future research). Another point of caution is that very high pulling velocities have been used; even though we have studied a range of deformation speeds in order to better understand the rate dependent behavior of these protein structures, physiological and experimental deformation rates are usually several orders of magnitude smaller.

Although an increasing number of mutations has been found in lamins, individual mechanism explaining how single mutations lead to various tissue-specific diseases are still unknown [156]. Muscles largely depend on proper mechanical properties to function correctly under physiologic conditions. Knowing how a single mutation causes impairment in mechanical properties of lamins will help us to understand more about the origin of laminopathies, and in particular lamin related muscle dystrophies. This knowledge could play a key role in developing potential therapies for laminopathies or even cancer [73, 156]. Moreover, it could increase our understanding of protein materials.

Silencing and activation through hierarchical scales

These observations suggest that the single missense mutation considered here does not affect the mechanical properties of lamin on the dimer level. This observation is backed up by the understanding of the origin of the stability and mechanical properties of an AH or a CC: According to previous simulation studies of CC structure [24], it was observed that the mechanical properties of the structure are primarily controlled by HBs. Although the mutation introduces a charge change in the side chain, the HBs that stabilize the AH structure within the CC are still intact, providing a reasonable explanation for the unchanged mechanical properties. This observation also justifies, why in the derived theory (Section 4) the primary structure was not considered as long as the secondary structure is known or not influenced by mutations.

Further, the mechanical behavior of the lamin dimer – featuring the three regimes during the tensile test – is very similar to the behavior of the vimentin dimer, as will be shown later. This suggests that nuclear IFs have qualitatively similar mechanical properties as cytoplasmic IFs, underlining the idea that the secondary and the tertiary structure, and much less the primary structure determine the mechanical properties of individual proteins. This is only true as long as the primary structure leads to formation of the particular secondary structure.

In a broader context, the observation of identical mechanical properties of wild type lamin compared to the mutated protein as well as in comparison with vimentin dimers, is an example for the concept of silencing and activation. Of course, the changed information on the lowest hierarchical scale (*i.e.* the AA sequence) is forwarded through hierarchical scales without visibility at intermediate scales to higher hierarchical scales, where these mutations are activated [157]. In this case, the point mutation does not affect the building of HBs and thus the geometry of the secondary structure. Contrarily, the change in charge could lead to a change in the tertiary structure or other structural and chemical changes at higher hierarchical scales (as will be shown in Section 5.4). For example, it was shown that mutations in desmin IFs do not influence the secondary structure (AHs) but lead to additional stutters or stammers (tertiary structure) [158]. The idea of silencing and activation is discussed in more detail in Chapter 7.2.2.

5.2 Studies of secondary structures: Deformation and fracture in AHs

In this case study, we focus on the deformation and fracture behavior of a single AH, over ten orders of magnitude time scale.

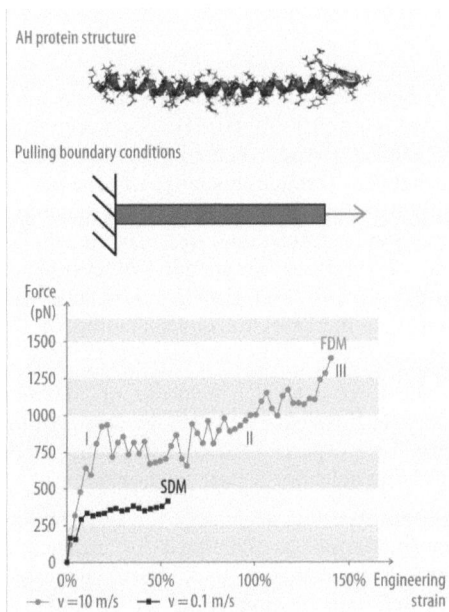

Figure 5.7: Characteristic force-strain curves of stretching single AHs in different deformation modes (depending on the pulling speed). The fast deformation mode (FDM) is represented by a curve taken from a pulling experiment with a pulling speed of 10 m/s. The slow deformation mode (SDM) is represented by a pulling experiment at 0.1 m/s. The general force-extension behavior consists of three regimes: The first regime shows a linear increase in strain until the angular point (AP) is reached. At this point, the first HB ruptures, leading to unfolding of one helical convolution. The second regime is a plateau of slightly increasing force, during which unfolding of the entire protein occurs. The slight increase in force appears due to the length effect, which was postulated in the theory (Equation (4.24)). The Decreasing remaining length and thus the number of serial convolution results in a logarithmic increases in force. The last regime (only partly shown for the FDM) displays significant strain hardening, leading to a progressive increase in stiffness. Due to computational limitations, the simulations of the SDM are carried out only up to 50% strain (the total simulation time in the SDM is 40 ns). Our results show that the unfolding in the SDM begins at approximately 10% strain, whereas in the FDM the unfolding begins at approximately 20% strain.

5.2.1 Protein structure

The AH motif is commonly found in structural protein networks and plays an important role in biophysical processes that involve mechanical signals, which include mechanosensation, mechanotransduction, as well in providing mechanical stability to cells under large deformation [15, 50]. For instance, AHs are part of the mechano-transduction IF networks that forward signals from the cellular environment to the DNA, where specific response signals are triggered [59, 60].

A detailed understanding of these transduction chains and the chemo-mechanical coupling are vital to gain insight into the cellular processes such as cell mitosis or apoptosis. Thus, the mechanical properties of AHs and the link to associated atomistic-scale chemical reactions are of vital importance in biophysics, biochemistry and biomedicine, as well as for the *de novo* design and manufacturing of nano-featured PMs [159-161], where AH motifs are utilized as a self-assembling building block for bottom-up designed biomaterials. Further details regarding the geometry of AHs were provided in Section 2.2.

Atomistic protein structure

The AH structure used in this study is taken from the 2B segment of the vimentin CC dimer of length 70 Å (with Protein Data Bank (PDB) ID 1gk6). As presented earlier, vimentin belongs to the group of IFs, which are – in addition to MFs and MTs – one of the three major components of the cytoskeleton in eukaryotic cells [33, 36, 57]. They are considered to play a crucial role in mechanical resistance of cells at large deformation (see also Section 2.4).

The mechanics of AHs

The mechanics of AHs plays a crucial role in biology, ranging through disparate time-scales reaching from picoseconds to seconds and more [10, 59, 60]. However, currently there exists no model that describes the mechanical strength behavior of AH protein domains that considers associated physical mechanisms through this range of time-scales; experiments have been carried out at relatively slow pulling rates, and computer simulations have been carried out at much faster deformation rates. The results of experiments and computational studies have not yet been integrated. However, such studies are critical as they do not only allow to link MD simulation results with experiments and with *in vivo* conditions, but also increase the understanding of the protein's behavior on different time scales.

Further, a structure-property relationship for the force-extension behavior of AHs and associated strength models has not been reported. No links exist between the details of the molecular architecture, the resulting free energy landscape and the resulting mechanical properties. The detailed mechanisms of the unfolding behavior remain unknown.

Thus far, different deformation modes have been observed separately for slow pulling rates in experiment and fast pulling rates in simulation, as illustrated for instance in references [162, 163]. Transitions of unfolding mechanisms have been suggested, however, never observed directly in either experiment or simulation. In particular, it remains controversial if the free energy landscape of the unfolding behavior of proteins consists of multiple, discrete transition states or if the transition states change continuously with a change in pulling velocity [162, 163].

However, this understanding is crucial to design new materials and tissues with multiple nanoscopic features. It is also vital to advance strength models of PMs, since macroscopic fracture is controlled by local protein unfolding at the atomistic and molecular scale. Such advances are critical on the path to understand the molecular mechanisms of the processes such as mechano-transduction [59, 60]

Here we resolve the issue of time scales and deformation mechanisms by providing a self-consistent approach that allows us to predict the strength of AHs over more than ten orders of magnitude in time scales, quantitatively linking atomistic simulation results with experimental results, based on fundamental physical parameters that include the energy and geometry of HBs and the persistence length of the protein's backbone. The model captures the behavior of AHs from "slow" natural biological processes up to mechanical shocks as they appear in accidents and injuries.

5.2.2 Mechanics at high and intermediate pulling rates

A series of MD simulations of a AH protein were carried out (Figure 5.7 (a)) with the CHARMM/NAMD (see Section 3.2.1 for details) method, using explicit solvent. The goal of this study is a systematic analysis of the unfolding behavior of the AH protein at different pulling rates.

Observation of two discrete energy barriers at intermediate and high pulling speeds

Two characteristic force-strain curves are shown in Figure 5.7, obtained for two pulling speeds. The simulations reveal - similar to the results shown for lamins – the existence of three distinct deformation regimes. The first regime shows a linear increase in strain until the AP is reached. The second regime is a plateau of slightly increasing force, during which unfolding of the entire protein occurs. The last regime displays a significant strain hardening, due to pulling of the protein's backbone (only partly visible in the FDM plot). The change from the first to the second regime is referred as the AP, and the force at this point is the protein unfolding force. The AP is significant since it represents the beginning of unfolding of the protein, characterized by rupture of the first HB(s) that consequently destroys the characteristic AH structure. This has been confirmed by an analysis of the atomistic geometry of the AH structure as the force is increased. Further, the AP determines the second regime, which governs over 100% strain in the force-strain curve. In the remainder of this Thesis, we focus on the force at the AP as a function of the pulling speed.

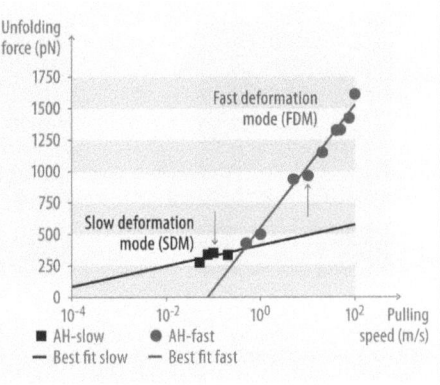

Figure 5.8: Unfolding force of single AHs from the vimentin CC dimer, as a function of varying pulling speed over four orders of magnitude, ranging from 0.05 m/s to 100 m/s. The arrows indicate the representative pulling speeds used for the analysis reported in the other figures. As predicted by Equation (4.18), the unfolding force of the AH depends logarithmically on the pulling speed. The results clearly reveal a change in mechanism from FDM to SDM at 0.4 m/s pulling speed, and at a force of approximately 350 pN. This suggests a free energy landscape that consists of two transition states for the structure studied here.

We carry out computational experiments by systematically varying the pulling velocity over four orders of magnitude, ranging from 0.05 m/s to 100 m/s. The unfolding force is plotted over the logarithm of the pulling speed in Figure 5.8. Notably, we observe two distinct regimes, each of which follows a logarithmic dependence of the unfolding force with respect to the pulling rate. The existence of two discrete slopes indicates existence of two different energy barriers and thus two different unfolding mechanisms over the simulated pulling velocity regime. In the following we refer to these two regimes as the slow deformation mode (SDM) and the fast deformation mode (FDM). The FDM appears at pulling velocities higher than the critical pulling speed $v_{cr} = 0.4$ m/s and a critical unfolding force $F_{cr} = 350$ pN. The SDM is observed at values smaller than v_{cr} and consequently, F_{cr}. This regime is more closely related to experimental or *in vivo* pulling velocities.

To the best of our knowledge, up until now neither any unfolding behavior in the SDM, nor the change from the FDM to the SDM was observed in direct MD simulation or in experiment. We emphasize that the change in mechanism has thus far only been suggested or inferred [163, 164]. For example, a comparison between MD simulation and experimental results revealed that force-pulling speed dependence must lay on two different curves in the f-log(v) plane [163, 164], suggesting a change in unfolding mechanism. However, this change in mechanism has never been directly confirmed.

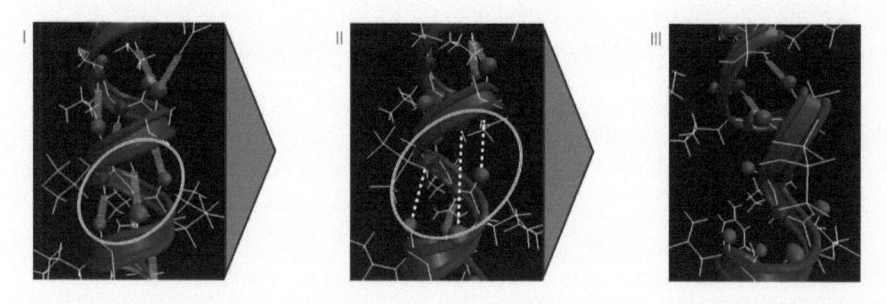

Figure 5.9: The unfolding of the AH in the SDM (AP in Figure 5.7, for $v = 0.1$ m/s) starts with simultaneous rupture of three HBs. The time interval between these snapshots is 20 ps (between I and II) and 40 ps (between II and III). After 20 ps (from I to II) all three HBs have ruptured simultaneously, leading to local unfolding of the protein in the following 40 ps (from II to III). We note that it was reported in the literature that the time for HB breaking is of approximately 20-40 ps [25]. Thus these snapshots strongly support the concept of cooperative bond rupture in the SDM. Surrounding water molecules are not shown for clarity.

SDM is governed by rupture of three parallel HBs

By fitting the Hierarchical Bell Model to MD results, we obtain for the FDM $E_b = 4.87$ kcal/mol and $x_b = 0.21$ Å. In the SDM, $E_b = 11.1$ kcal/mol and $x_b = 1.2$ Å. The value for the angle θ was measured with VMD to 16°. Considering that the bond breaking energy E_b^0 of a HB in water ranges from 3 to 6 kcal/mol [25], this indicates that in the FDM, one HB ruptures at once. In contrast, in the SDM approximately three HBs rupture at once. Consequently, by inserting these numbers in Equation (4.17), we calculate $E_b^0(SDM) = 4.21$ kcal/mol and $E_b^0(FDM) = 5.83$ kcal/mol. The value $E_b^0(FDM)$ is slightly higher due to the appearing length effect (see Equation (4.24)), as having several bonds in sequence reduces the effective energy barrier.

The smaller x_b in the FDM than in the SDM (see Table 5.1) is due to strain localization in the molecule, that is, the local strain is several times larger than the laterally applied strain (see Figure 5.10 for details). This amplification of local strain leads to a smaller value of x_b. Further, before the first HB ruptures at the AP, all HBs in the molecule are stretched by approximately 0.21 Å, which equals to 7% HB strain (equilibrium HB length is 3 Å). At the AP, one of the HBs rapidly extends to $x_b = 1.2$ Å and ruptures. In contrast, in the SDM the protein is not homogeneously stretched in the first regime. Rather, one random convolution is stretched so that 3-4 HBs extend homogeneously. Subsequently, three HBs rupture simultaneously, as shown in Figure 5.9. The mechanism is shown schematically in Figure 5.10, which plots the microscopic strain distribution before and after the AP is reached, comparing the FDM and the SDM regime.

A detailed analysis of the atomistic structure during HB rupture is shown in Figure 5.9 for the SDM. This analysis provides additional evidence that in the SDM, three HBs rupture simultaneously, within less than 20 ps time scale. We note that it was reported that the time for HB breaking is of approximately 20-40 ps [25], clearly supporting the notion that these HBs rupture at once. We also can calculate this value directly by taking Equation (4.3), a $E_b^0 = 4$ kcal/mol and $F = 0$.

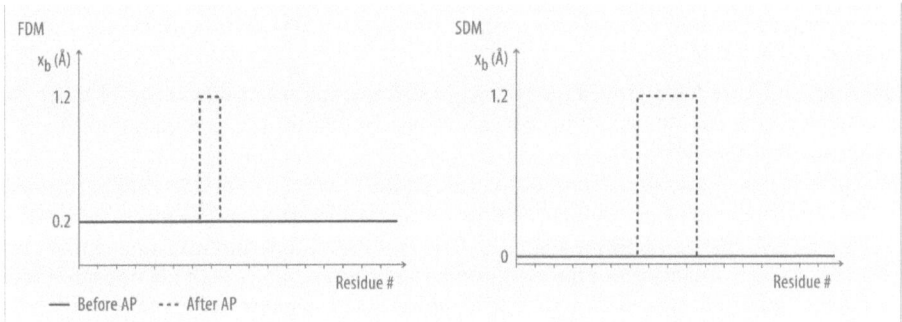

Figure 5.10: The microscopic strain distribution before and after the AP is reached, comparing the FDM and the SDM regime. We find that in the FDM all HBs in the molecule are homogeneously stretched by approximately 0.2 Å, before the first HB ruptures at the AP. At the AP, one of the HBs then rapidly extends to x_b=1.2 Å and ruptures. An analysis of the molecular strain distribution reveals that this represents a localization of strain. In contrast, in the SDM, the entire protein is not homogeneously stretched in the first regime. Rather, one random convolution is stretched so that 3-4 HBs extend homogeneously. Then, three HBs rupture simultaneously, as shown in Figure 5.9.

Further evidence for the change in mechanism is obtained by an analysis of the HB rupture dynamics. In Figure 5.11 we plot the HB rupture as a function of the molecular strain ε. This provides a strategy to normalize the different time scales by the pulling velocity (here 0.1 m/s and 10 m/s). In agreement with the results shown in Figure 5.7, the unfolding of the protein in the SDM starts at approximately 10% strain, in contrast to 20% strain in the FDM regime. This is indicated in Figure 5.11 by the rupture of the first HB. We clearly observe in Figure 5.11 that in the FDM, HBs rupture sequentially as the lateral load is increased from 20% to 40% tensile strain. In contrast, in the SDM several HBs rupture virtually simultaneously, within less than 20 ps, at a tensile strain of ≈ 10%. Even though the pulling speed is several orders of magnitude slower in the SDM, the time in which HBs in the SDM rupture is significantly shorter.

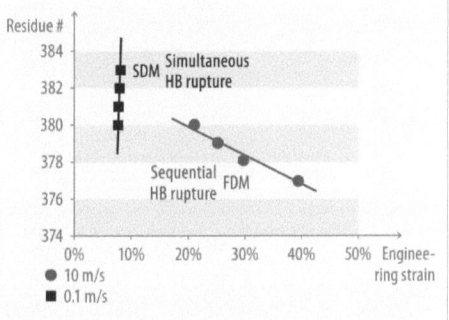

Figure 5.11: The rupture sequence of the first four HBs, which represent one convolution as a function of the applied strain. The residue number represents the amino acid of the O-atom (hydrogen acceptor). We can clearly see that in the FDM the HBs rupture one by one, whereas in the SDM several HB rupture virtually simultaneously, within 20 ps. Even though the pulling speed is several orders of magnitude slower, the HBs in the SDM rupture significantly faster. As shown in Figure 5.7, the unfolding in the SDM regime begins at 10% strain in contrast to the 20% strain in the FDM. We note that in the FDM the unfolding wave runs from the pulled residue in the direction of the fixed residue, whereas in the SDM the unfolding 'wave' runs in the opposite direction, nucleating at a random residue within the protein sequence.

The details of the atomistic rupture mechanisms in the FDM and the SDM regimes are summarized in Table 5.1.

Unsurprisingly, we have proved the change in deformation mechanism also for other AH structures, *i.e.* a AH domain from bacteriophage T4 fibritin [165], suggesting this is a universal feature of AHs[26].

We use VMD for visualization of protein structures [101], as well as for the analysis of the length of HBs over the simulation time (Figure 5.11). The rupture length of a HB is defined as 5 Å (the equilibrium length of HBs is ≈ 3 Å, [166] and measurements in our simulation). The distance from the equilibrium was chosen higher than the transition state as known from theory and experiment in order not to capture the dynamics of bond breaking and rebinding due to thermal fluctuation.

In previous atomistic simulations, unfolding forces have been significantly larger than those measured in experiment, likely because they were carried out in the FDM so that forces reach up to several nN for individual AHs. This is clearly an artifact of excessively

Property	Deformation mechanism	
	Slow deformation mode (SDM)	Fast deformation mode (FDM)
Pulling speeds (m/s)	$v_{cr} < 0.4$	$v_{cr} > 0.4$
Unfolding forces (pN)	F < 350	F > 350
E_b (kcal/mol)	11.1	4.87
x_b (Å)	1.2	0.21
HB breaking mechanism	Simultaneous rupture	Sequential rupture

Table 5.1: Summary of the major differences between the SDM that governs the behavior at pulling velocities below 0.4 m/s and 350 pN, and the FDM that governs the unfolding behavior at larger pulling velocities

large pulling speeds [162, 163]. Our analysis shows that in addition to incorrect force estimates, the observed unfolding mechanism can also be significantly different if the pulling speed is too high. The estimate for v_{cr} provides a 'maximum' pulling rate that could be used in MD studies, in order to still allow a reasonable interpretation of MD results in light of biological relevance. The quantitative values derived here may therefore provide guidance to set up other MD simulations. The SDM is more relevant for biological function. However, the FDM could be important during tissue injuries that may be incurred under large deformation rates (e.g. shock impact or fractures and trauma).

5.2.3 Mechanics at ultra small pulling rates

As shown in previous Section the predictions from the Hierarchical Bell Model lead at decreasing loading rates even in the SDM to negative forces (at pulling speeds bellow v_0), a rather unphysical prediction. Furthermore, experimental values [167, 168] clearly do not lie on an extension of the slope predicted from the SDM regime, and rather suggest that the f-log(v) curve approaches an asymptotic zero slope (see Figure 5.13 (a)).

Could the Bell model be used to explain this behavior at vanishing pulling rates? Adopting the Bell model to describe this observation would lead to an increase of x_B (since x_B controls the slope of the f-ln(τ) or the f-ln(v) curve), approaching infinity for slopes approaching zero. This would also be unphysical since the transition point x_B

can not be larger than the finite contour length of the protein domain. This suggests that another mechanism must control the strength.

The key to understand this change in mechanism is the realization that at sufficiently long time scales the deformation of the system goes through equilibrium and is no longer controlled by a statistically activated process as described by the Hierarchical Bell Model (see Equation 4.3). Thus the strength does not depend on the time-scale beyond a critical τ_{eq} or pulling speeds lower than v_{eq}. Thereby v_{eq} is slightly higher than v_0, as the asymptotic regime (AR) is reached at forces higher than 0. We note that some of the models reported in the literature [40, 126, 129, 131-134, 141, 147] predict a continuous change in the slope of the f-$\log(v)$ curve; however these models contain parameters that can not be linked directly to distinct physical mechanisms. Here we provide an alternative explanation.

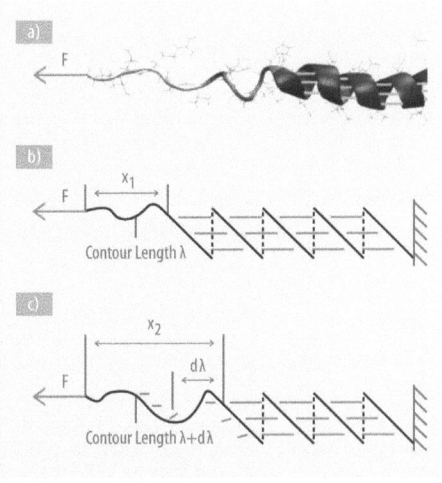

Figure 5.12: Subplot (a) depicts the atomistic-scale protein structure of a single alpha helix (AH) from a vimentin CC dimer. The helical backbone is stabilized by parallel arrangements of HBs (yellow dashed lines). Subplots (b) and (c) show a schematic model system of an AH strained by an external force before and after onset of rupture, showing the process of releasing a segment of backbone polypeptide due to the rupture of HBs, thereby increasing the contour length of the free end entropic chain by $d\lambda$. The ration of x/λ is defined as α.

At long time scales $\tau > \tau_{eq}$ entropic effects that stem from conformational changes of the polypeptide chain are activated and the strength is characterized by a free energy release rate condition, as recently reported in [152] for another class of protein domains (see Figure 5.12 for the application to AH structures). Here we apply this model [152] to AH protein domains. Similar to the Griffith condition used to predict the onset of fracture in crystals (an engineering approach) [169], the free energy released by freeing polypeptide chains from their geometric confinement in helical convolutions, must equal the energy required to break these HBs. The strength of the AH protein domain is then given by

$$f_{AR} = \frac{k_B T}{4\xi_P}\left[(1-\alpha_{cr})^{-2} + 4\alpha_{cr} - 1\right], \qquad (5.1)$$

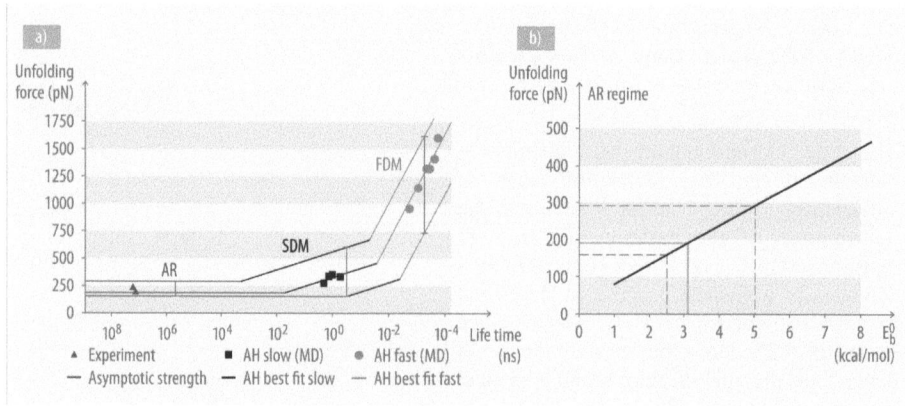

Figure 5.13: Subplot (a) shows the rupture force versus life time of the AH-system at the onset of failure (=strength properties), including all three regimes over more than ten orders of magnitude of time-scales. MD simulation results suggest a change in mechanism from the FDM to the SDM at increasing the time scales. At approximately 350 pN the effective energy barriers under the applied force in the Bell model are comparable, and therefore mark the transition between FDM and SDM mechanisms. At longer time-scales there is another change in deformation mechanism from the SDM to the asymptotic regime (AR), predicted here at a time scale of approximately 100 ns when $f_{AR} > f_{SDM}$. Experimental results confirm this prediction. Thin lines show the strength behavior for a broad range of HB energy values from 2.5 kcal/mol to 5 kcal/mol (marking error bars for uncertainties in the HB energy). Subplot (b): Dependence of the critical rupture force on E_b^0, in the AR. The strength of the system near equilibrium conditions (AR) depends linearly on E_b^0 (this parameter determines the energy release rate γ). The specific value of E_b^0 is usually found in a range between 1 and 8 kcal/mol, and varies between different solvent conditions and the specific sequence of the protein domain.

with α_{cr} obtained from

$$G(\alpha_{cr}) = \frac{k_B T}{4\xi_P}\left[\alpha_{cr} \cdot (1-\alpha_{cr})^{-2} - (1-\alpha_{cr})^{-1} + 2\alpha_{cr}^2 + 1\right] \stackrel{!}{=} \gamma. \qquad (5.2)$$

Hereby α equals to the ratio of the end-to-end length of the free chain to its contour length $\alpha = x/\lambda$ (see Figure 5.12 for definition of variables), equivalent to mechanical stretch. The parameter γ describes the HB energy stored per unit length of AH,

$$\gamma = \frac{E_b^0 \cdot n}{l_0} = \frac{E_b^0}{L_0}, \qquad (5.3)$$

With l_0 as the unit length of one convolution and n as the number of HBs per unit length. We calculate $L_0 = 0.145$ nm/HB by measuring the length of the entire protein ($L_x = 6.9$ nm) and dividing it by the number of existing HBs ($n = 47$). This is in good agreement with results in the literature [50] where $L_0 = 0.15$ nm/HB (calculated from $l_0 = 0.54$ nm and $n_0 = 3.6$ HBs per convolution). The strength regime described by Equation (5.1) is referred to as asymptotic regime (AR).

Combing all three mechanisms (FDM, SDM, AR), the strength of an AH domain is:

$$F(\tau; x_b^{FDM}, E_b^{FDM}, x_b^{SDM}, E_b^{SDM}, \theta, \xi_p, \gamma) = \max \begin{Bmatrix} f_{FDM}(\tau; x_b^{FDM}, E_b^{FDM}, \theta) \\ f_{SDM}(\tau; x_b^{SDM}, E_b^{SDM}, \theta) \\ f_{AR}(\xi_p, \gamma(E_b^0)) \end{Bmatrix}. \quad (5.4)$$

The functions f_{FMD} and f_{SMD} can be calculated from Equation (4.16) or (4.24), f_{AR} can be calculated from Equation (5.1-5.3). We estimate E_b^0 from the MD simulation in the FDM where the 3.6 HBs in one convolution break simultaneously, thus $E_b^0 = E_b^{SDM}/3.6 = 3.1$ kcal/mol, thus $\gamma = 2.1$ kcal/mol/Å (this provides a direct link between SDM and AR). Alternatively, we could take an $E_b^0 = 4.21$ kcal/mol as calculated in Section 5.2.2. This value of E_b^0 is in good agreement with earlier experimental and simulation results [25], where E_b^0 was reported to be 3-6 kcal/mol. We choose the persistence length of a polypeptide chain to $\xi_p = 4$ Å (a widely accepted parameter that has been measured in experiment and confirmed in simulation studies) [170]. Based solely on these two parameters, E_b^0 and ξ_p, the force in the AR is calculated to ≈190 pN. The AR regime is reached at time scales of $\tau_{eq} \sim 100$ ns (or equivalently, at pulling speeds $v_{eq} < 0.001$ m/sec), when $f_{AR} > f_{SDM}$. The strength value of f_{AR} is plotted in Figure 5.13 (b) as a function of the HB energy E_b^0.

The model given in equation (5.4) is validated through quantitative comparison with experimental results. Experimental results of stretching and breaking single AH domains [167, 168] (with a length of less than 100 Å) report forces between 140 and 240 pN during unfolding. Figure 5.13 (a) summarizes the described regimes and shows a quantitative comparison between the model prediction and MD simulation results as well as experimental results. In addition to the values used in this study that were based on earlier MD results, an envelope curve for E_b^0 ranging from 2.5 kcal/mol to 5 kcal/mol is

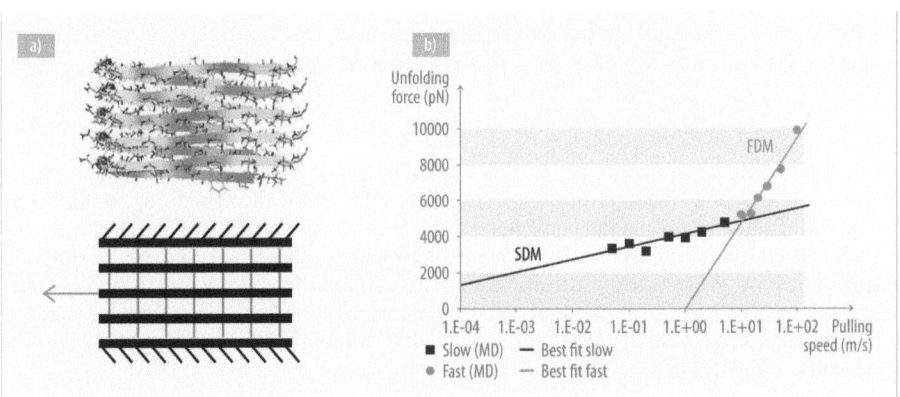

Figure 5.14: Subplot (a): Atomistic geometries of the BS studied here. Surrounding water molecules are not shown for clarity. The lower part of the plot indicates the boundary conditions (shear loading). The BS structure consists of two stacks of beta-sheets in the out-of-plane direction. Subplot (b): Unfolding force of a BS amyloid domain as a function of varying pulling speed over four orders of magnitude, ranging from 0.05 m/s to 100 m/s. The results clearly reveal a change in protein unfolding mechanism from the FDM to the SDM. Data generated by Sinan Keten and Xuefeng Chen.

included to illustrate how the predictions change under variations of the energy of HBs. We note that other experimental results [170-174] (not shown in Figure 5.13 (a)) that consider AH spectrin repeats lay slightly below the predicted force range, on the order of 50 pN, which would require extremely low values of $E_b^0 \approx 1$ kcal/mol. A possible explanation for this behavior could be a difference in the observed unfolding mechanism, which is the unfolding of the anti-parallel CC repeat instead of rupture of individual HBs of a AH domain. There is evidence for this, since in one of the studies x_b was estimated to be 15 Å [170-174], which is ten times higher than the x_b for a single HB thus suggesting an alternative mechanism.

5.2.4 Change in deformation mode for beta-sheets

In the last sections we have discussed the change in deformation mechanism for AH protein domains. To understand whether or not this is a phenomenon that is also relevant for other protein structures, we present atomistic simulations carried out on beta sheets (BS, schematic in Figure 5.14 (a) upper part; results generated by Sinan Keten and Xuefeng Chen [26]).

The BS model we use in this study is a protein structure proposed for Alzheimer's amyloid β-fibril [175]. As BS mainly underlie to shear load, we pull on the middle chain of the assembly (third chain from top or bottom) at the midpoint of the turn that connects the two beta strands. We fix all of C_α atoms on top and bottom chains during pulling (Fig. 5.14(a), lower part). These boundary conditions are similar to those reported in recent AFM experiments of a similar amyloid structure [176].

Figure 5.14 (b) shows the unfolding force of the BS domain as a function of the pulling speed. Notably, also in this case we observe two distinct regimes, each of which follows a logarithmic dependence of the unfolding force with respect to the pulling rate. The existence of two discrete slopes indicates two different energy barriers and thus two different unfolding mechanisms over the simulated pulling velocity regime. The results clearly suggest a free energy landscape that consists of two transition states. For the BS structure, the transition occurs at v=10 m/s at a force of 4,800 pN.

We obtain from our model for the FDM E_b = 2.2 kcal/mol and x_b = 0.024 Å. In the SDM, E_b = 11.1 kcal/mol and x_b = 0.138 Å (the angular term θ is not considered in this case). Notably, the force levels in the BS domain are much higher than in the AH structure, indicating that this protein domain may be mechanically sturdy, approaching rupture forces at experimental and physiological pulling rates of 1 nN. These strength values agree qualitatively with recent experimental studies of amyloid structures, explaining how shear loading of arrays of HBs can lead to extremely strong resistance against rupture [176, 177]. Similar to AHs, an analysis of the bond breaking history revealed that sequential breaking of single HBs occurs in the FDM, and concurrent breaking of ≈ 6 HBs in the SDM.

Notably, the asymptotic regime, which appears at vanishing pulling rates and time scale approaching the equilibrium was also reported for BS [152].

5.2.5 Conclusions in the light of materials science and biological function

The fact that the change in deformation mechanism from simultaneous rupture of HBs to sequential rupture with increasing pulling speeds is observed for three protein structures under different loading conditions (tensile loading for the AH domains, and shear loading for the larger BS domain) suggests that the discrete change in mechanism may be a universal phenomenon. In particular, the results obtained from the BS structure illustrate that this transition appears also at larger hierarchical levels indicating the broadness of the phenomenon.

We note that the interface of different proteins or even the supermolecular structure is very significant and may be most relevant for many biological functions (for instance, the unfolding of globular domains in titin or unfolding of spectrin at the linker region between two alpha-helices under strain [163]). However, in order to predict the deformation mechanisms of more complex protein structures, studies as the one reported here are critical as they enable one to compare the strength of different competing deformation modes. (Larger-scale protein networks will be investigated in Section 6.2 to address some of these issues.)

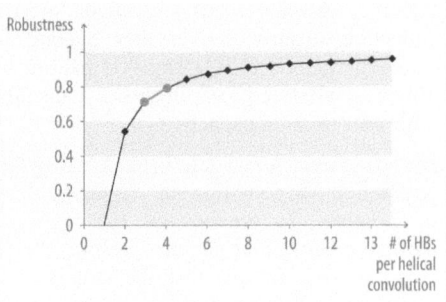

Figure 5.15: Robustness as a function of parallel HBs per convolution (parameter b_1). Robustness is defined as the ratio of strength of a failed system and an intact system. The intact system is defined as a system where all HBs contribute to strength, whereas in the failed system all except one HB contribute to the strength of the AH. A robustness of "0%" means that the system is fragile, whereas robustness of "100%" represents a fault tolerant material. As can be seen in the graph, the robustness converges towards fault tolerance when b_1 goes to infinity. The round data points (in orange) indicate the number of parallel HBs per convolution (3.6 HBs) as it is found in Nature. It is observed that this particular molecular geometry corresponds to a robustness value of $\approx 80\%$, indicating that the AH is efficient in Pareto's sense [16-18]. Robustness is calculated following Equation (5.5).

Another important contribution of this Chapter is the development of a constitutive model that describes the strength properties of AH protein domains over more than 10 orders of magnitudes of time-scales. Up until now such a model has not been reported, and to the best of our knowledge this model is the first to quantitatively link MD simulation results [26] and experimental AH strength values [167, 168] in a simple physical model as shown in Figure 5.13. An important feature of the model reported in Equation (5.4) is that it only includes basic parameters of the protein structure, that is, the HB energy and geometry, as well as the persistence length. It is found that the strength properties of the AH protein domain, a universally found biological protein structure, is controlled by different mechanisms at distinct time scales, with strong strengthening under faster rates (shorter time-scales). Our study could motivate new experiments, in particular at higher deformation rates to probe the transitions between the regimes described here. Other approaches may utilize mesoscale modelling methods to investigate the behavior of AH protein domains at larger ranges of time-scales.

5.2.6 AHs may follow Pareto principle in maximizing robustness

Protein folding and thus the generation of hierarchical structures are essential for biological function. First, it allows distant parts of the AA chain to come physically closer together, creating local sites with specific chemical properties that derive from the collection of particular residues. Second, folding permits collective, localized motion of different regions [178]. The AH pattern is the most simple folding motif out of a one dimensional strand [50], forming a spring-like protein structure with high elasticity and large deformation capacity.

But why does an AH fold in such a way that 3.6 parallel HBs, instead of 2, 5, or 6 appear in parallel, per convolution? Notably, all AHs universally show this particular molecular architecture. To the best of our knowledge, there has been no explanation for this particular molecular feature, despite the fact that the AH is such an abundant protein structure. We believe that the structural features can be explained by considering the robustness of the AH structure against mechanical and thermodynamic unfolding.

We calculate robustness based on the definition of robustness as parameter insensitivity, postulated by Kitamo [179, 180]. This definition applied to the case of an AH structure corresponds to the sensitivity of the protein strength in regards to missing HBs. Starting with the Hierarchical Bell Model (Equation 4.18), we calculate robustness as the ratio of strength of a defected system and an intact system. The intact system is defined as a system where all HBs contribute to strength, whereas in the defected system all except one HB contribute to the strength of an AH. Only the contributions due to the hierarchy in Equation (4.18) are considered; the pulling speed part of this equation is not taken into account since we compare systems at identical pulling speeds. The robustness is then defined as

$$r(b_1) = \frac{f_{h1}(k_1 = b_1 - 1) + f_{h0}(k_1 = b_1 - 1)}{f_{h1}(k_1 = b_1) + f_{h0}(k_1 = b_1)} = 1 + \frac{k_B \cdot T \cdot \ln(b_1) - E_b^0}{E_b^0 \cdot b_1} \tag{5.5}$$

It can be seen from Equation (5.5) that robustness converges towards complete fault tolerance when b_1 reaches infinity.

Figure 5.15 depicts the robustness of an AH as a function of parallel HBs per convolution. The analysis shows that for a AH structure with 3-4 HBs per convolution, 80% robustness is achieved (0% robustness means that the system is highly fragile towards catastrophic failure, and 100% represent a fault tolerant material).

This result provides a strong indication that AHs are efficient following Pareto's principle [16-18], also known as the 80/20 rule. Pareto's efficiency rule found broad acceptance and application in explaining many social, economic, political, and natural phenomena. Our results indicate that this concept may be also applicable to explain the architecture of the AH protein motif. The more robust the structure becomes with each additional HB, the higher is the barrier to implement an additional HB, since each HB introduces an additional 'cost' due to increased material use that is, the additionally generated weight and additionally required volume. We emphasize that this 'implementation barrier' is the theoretical reason for the appearance of the Pareto distribution in other systems (see below for additional information on Pareto's principle).

In light of these considerations, it is not surprising that a robustness value of 80% is found in AHs, which equals the optimal state due to these concurrent mechanisms. This level of robustness in a biological structure enables to minimize waste of resources (AAs), the weight and volume, and in that way making the structure overall efficient and enables to sustain extreme mechanical conditions (such as high loading rates and deformations). This leads to an optimal combination of strength and robustness.

This finding is significant since the only input parameters in this model is the dissociation energy of a HB, E_b^0. This parameter can be determined reliably from either experiment or atomistic simulation (both approaches lead to similar values). The remainder of the parameters required to predict the robustness properties can be derived from the geometry of the protein structure. The fact that the results of this model only depend on very fundamental, 'first-principles' properties of protein structures illustrates the significance of our finding.

Thermal stability of AHs and comparison with synthetic materials

Forming 3-4 HBs in parallel instead of forming a single, much stronger bond is also energetically favorable, in particular in light of the moderate assembly temperatures present *in vivo*. However, this only makes sense if three HBs rupture simultaneously so that they can provide significant mechanical and thermodynamic resistance against unfolding. This has indeed been shown to be the case at physiological strain rates. This suggests that this number of HBs is not only efficiently robust, but also most easy to create through self-assembly. This illustrates the intimate connection of structural properties to assembly and functional processes in PMs, as it will be discussed in detail in Section 7.2.2 of this Thesis.

Synthetic materials typically do not have such high levels of robustness and fault tolerance. This makes it necessary to introduce safety factors that guarantee the structure's functionality even under extreme conditions. For instance, an engineering structure like a bridge must be able to withstand loads that are ten times higher than the usual load, even if this load will never appear globally. However, this safety factor is necessary since these structures are very fragile due to their extremely high sensitivity to material instabilities such as cracks, which might lead to such high local stresses. However, if a crack does not appear during operation, 90% of the material is wasted. This shows the potential of engineering bio-inspired robust and efficient structures. The key may be to include multiple hierarchies and an optimal degree of redundancies, as illustrated here for the AH structure. A detailed discussion on bioinspired materials will follow in Section 7.3

Background information on Pareto principle

Pareto's principle of differentiating the 'significant few' from the 'trivial many' holds true in many social, economic, political, and natural phenomena. For example 80% of the wealth are concentrated on 20% of the entire population [181]. Similarly, 80% of the revenues, software usage or published papers typically stem from 20% of all customers, programmed code or authors [181]. The 80/20 rule is an empirical law that is found in many natural phenomena.

The theoretical foundation of this principle was reported by Chen et al. [17, 18]. They state that this particular distribution is the result of the probability of a new entry, which quantifies the height of the entry barrier. Most importantly, it was shown that the probability of a new entry has an inverse relation with the level of concentration of the 'significant few' (e.g. revenue concentration, which equals to revenues per costumer, or paper concentration, which equals to paper per publisher.) This leads to a concurrence between existing entities and the addition of new entities, leading to the characteristic

80/20 distribution. In our case we refer to the robustness concentration, the value of robustness divided by the number of HBs, which is related to incremental increase in robustness with each additional HB.

This concept can be applied to explain the particular molecular features of the AH structure: The more robust the structure becomes with each additional HB (see Figure 5.15), the higher is the barrier to implement an additional HB, since each HB introduces an additional 'cost' due to increased material use, that is, the additionally generated weight and additionally required volume. However, the increase of robustness decreases rapidly, so that an optimal balance is found at the characteristic 80% mark. Our analysis therefore suggests that the Pareto rule may also be relevant in explaining the structure of AH proteins in light of their robustness. To the best of our knowledge this is the first time that the Pareto concept has been applied to explain the structure of proteins.

5.3 Studies of tertiary structures: Defects in CCs

In the previous case study reported in Section 5.2 [24] we have carefully examined the mechanics of a regular AH protein domain. Here we focus on the mechanics of a CC protein with a stutter defect (for definition see Section 2) and compare its mechanical behavior with a regular CC structure.

5.3.1 Protein structure

This study involves a systematic analysis of three molecular geometries: Model A is a 'perfect' CC *without* a stutter (PDB ID 1gk6, residues 355-406). Model B is a system that contains two single AHs (extracted from model A, entire structure embedded in explicit solvent), arranged in parallel with a distance of 4 nm, mimicking the molecular geometry in the stutter region. Model C is a CC structure *with* a stutter (PDB ID 1gk4, residues 328-379) and thus a 'combination' of model A and B. All structures have the same length of 70 Å, to exclude any effects of the molecular length on the results. Figure 2.2 depicts the location of models A and B in the vimentin dimer.

The strategy of investigation is as follows. We start with an analysis of models A and B. For both systems, we present an analysis using the Hierarchical Bell Model and confirm the theoretical predictions by MD simulation of both systems. Then, we consider model C and carry out MD simulations to analyze the unfolding dynamics and unfolding resistance.

5.3.2 Results of molecular modeling

Prediction by theory

Geometrically, the difference of model A and B/C is the angle θ, which describes the deviation of the HB direction from the pulling diretion (see also Figure 4.2). In model A, the angle $\theta_A = 23° \pm 10.2°$, whereas in models B and C, the angles $\theta_B = 16° \pm 8.5°$ and $\theta_C = 16° \pm 6.9°$. The angle fluctuates randomly over the different HBs. Note that in model B/C, the angle $\theta \neq 0°$ since in an AH each HB connects residue i with residue $i+4$, whereas a convolution consists of only ≈ 3.6 residues. Consequently, the HBs of a straight AH are slightly tilted in the direction of the twisting backbone. Thus the difference in the angle in model A and model B are due to the molecular twist in CC systems that appear in addition to the tilt of the HBs.

Provided that x_b and E_b are equal – that is, assuming the same unfolding mechanism – the Hierarchical Bell Model predicts that the structure with the larger angle is stronger (for $\theta_2 > \theta_1$, $\cos\theta_2 < \cos\theta_1$ or $\cos\theta_1 / \cos\theta_2 > 1$):

$$F_2(v) = \frac{\cos\theta_1}{\cos\theta_2} \cdot F_1(v).$$

Therefore, in a system that consists of a hybrid combination of models A and B, the location that corresponds to model B represents the weakest segment, which will unfold most easily under increasing applied tensile load.

This situation is resembled in vimentin IF, as represented in model C (see Figure 2.2). In this sense, introducing the stutter corresponds to the deliberate addition of a weakening 'defect' into the molecular structure. This suggests that the role of the stutter may be to provide a predefined and

MD results	Model A Perfect CC	Model B Aligned AHs	Model C CC with stutter
E_b (kcal/mol)	5.59	5.67	5.01
x_b (Å)	0.19	0.20	0.19
θ (degrees)	23± 10.2	16± 8.5	16± 6.9

Table 5.2: The parameters E_b, x_b and θ for the three models studied here. We note that while E_b and x_b direct results of MD simulation, the angle θ is determined from the molecular geometry

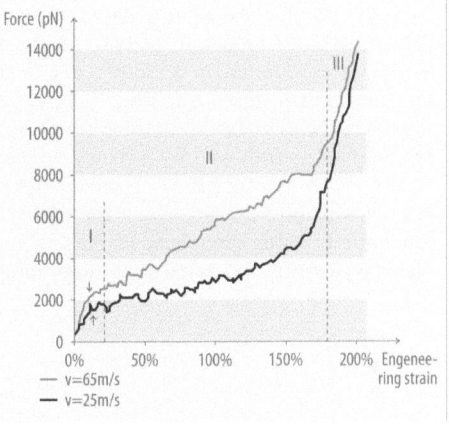

Figure 5.16: Force-strain curves of model A (CC AH structure) at two different pulling rates (the results of models B and C are qualitatively identical). One can distinguish three regimes: The first regime (I) represents elastic deformation, up to approximately 15 % tensile strain. This regime is followed by a plateau region (regime II) during which unfolding of the CC structure occurs, at approximately constant force level. The last regime displays significant strain hardening (regime III), during which the coiled super-helix is lost and the protein backbone is being stretched, leading to a significant increase in stiffness. The arrow in each curve marks the point where unfolding starts, the AP.

controlled location where unfolding occurs under large deformation.

For an angle $\theta_A \approx 23°$ of the CC (model A) and the angle $\theta_B \approx 16°$, the difference in unfolding force is approximately 5%.

As pointed out in Section 2, vimentin IFs have their most significant role in the large deformation behavior. Predicting the force at the AP and the level of force in regime II (Figure 5.16) is crucial, since these regimes dominate the large deformation behavior under physiological conditions.

Results of molecular simulations

First, we carry out a series of MD simulations with varying pulling rates, for models A and B, and measure the unfolding force as a function of pulling speed.

Figure 5.16 depicts representative force-strain curves of model A, at two different pulling rates. Similar to a simple AH, one can distinguish three regimes. The first regime represents elastic deformation, reaching up to approximately 10% tensile strain. This regime is followed by a plateau region during which unfolding of the CC structure occurs, at an approximately constant force level. The last regime displays a significant strain hardening, during which the coiled super-helical structure is lost and the protein backbone is being stretched, leading to a significant increase in stiffness. The arrow in each curve marks the point where unfolding starts.

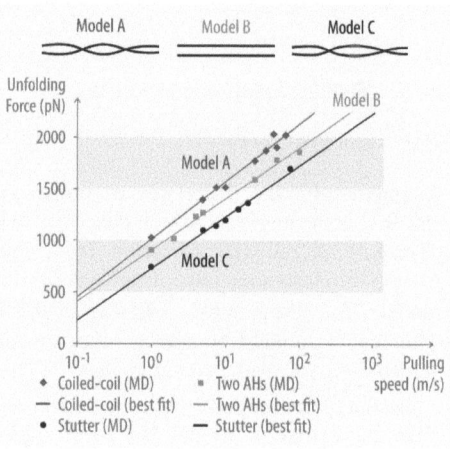

Figure 5.17: Unfolding force (at AP), for the three models, as a function of pulling velocity. All three models display a strong dependence of the unfolding force on the pulling velocity. Model A (perfect CC) shows the largest unfolding resistance, model B (two parallel AHs) has a lower strength. Model C (CC with stutter) has the smallest unfolding resistance. The straight lines represent best fits to the MD results.

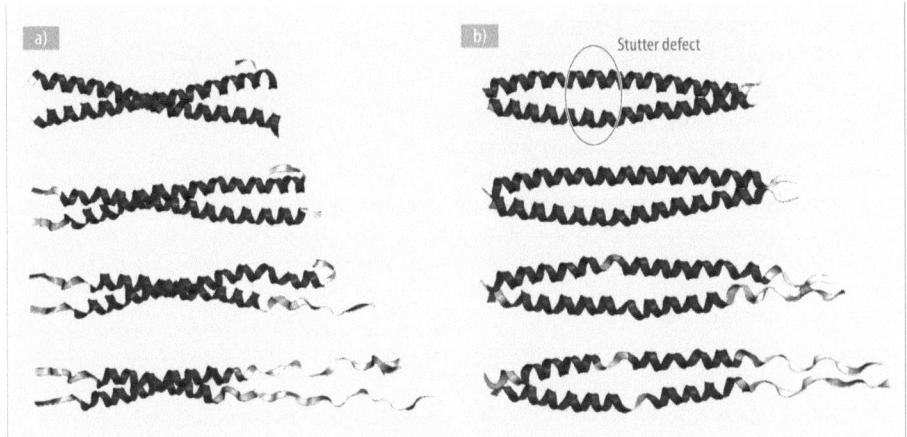

Figure 5.18: Subplot (a): Snapshots of the unfolding dynamics, model A (perfect CC). Unfolding proceeds sequentially, beginning from the end where load is applied. Subplot (b): Snapshots of the unfolding dynamics, model C (CC with stutter). The marker indicates the location of the stutter. It is apparent that the stutter represents a location where unfolding starts as well (see the 2nd and 3rd snapshot from the top, where it is evident that the AH motif begins to disappear at the location of the stutter). The structure is rendered using the VMD ribbons method; color is assigned according to the secondary protein structure. The initial length of the structure is 70 Å.

In order to gain more insight into the molecular unfolding mechanism as well as the rate dependence of the results, we have performed simulations over a wide range of pulling velocities. The unfolding force is shown in Figure 5.17 as a function of pulling velocity. In our analysis, the unfolding force was defined as the maximum force in the transition from the first to the second regime. One can clearly see that – in agreement with the predictions by the theory – model B and C features a reduced unfolding force compared with model A. Thus the molecular simulations clearly confirm the hypotheses put forward in the previous Section.

Further, fitting the parameters x_b and E_b directly from MD simulation results in Equation (4.4) with (4.5) reveals that they are almost identical, suggesting the same unfolding mechanism (see Table 5.2 for a summary). This provides strong evidence that the strengthening effect is due to the variations in the angle θ. The particular value of $E_b \approx 5$ kcal/mol suggests that the deformation mechanism is characterized by rupture of a single HB, in agreement with earlier studies [24].

For model C, the unfolding force is even lower than in model A and B. In models A and B, unfolding begins only at the ends of the molecule, most probably due to boundary effects. Most importantly, in model C, unfolding occurs also at the stutter, in agreement with the theoretical prediction. This is confirmed by the snapshots of the unfolding dynamics for both model A (Figure 5.18 (a)) and model C (Figure 5.18 (b)). We have further confirmed that the unfolding behavior is independent of the direction of pulling (results not shown).

5.3.3 Conclusions in light of materials science and biological function

As shown above, the stutter is a defect that can only appear in the CC geometry. Our model suggests that single alpha-helices with and without the stutter sequence will have similar unfolding behavior. This is because the weakening effect is due to the variation of the angle θ. However, the variation in the angle θ is not 'visible' at the scale of a single AH.

This study illustrates similarly to the AH case study in Chapter 5.2 the significance and role of utilizing hierarchies in the design of biological materials. Modifications in the lower hierarchy (AA sequence) do not influence the immediately following hierarchical level (AHs), but show effects just one hierarchical level higher, that is, on the CC level, where it has a profound impact on the mechanic properties. This illustrates that the impact of a small-scale feature can be silenced at an intermediate scale and become active at a larger scale. Similar effects have been already reported with studies on protein mutations in Section 5.1. This underlines our hypothesis of silencing and activation, which will be discussed in details in Chapter 7.2.2.

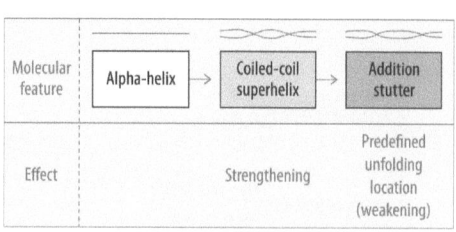

Figure 5.19: Summary of the effect of different molecular features. Formation of a CC super helix out of individual AHs leads to a strengthening effect. Addition of the stutter provides a molecular defect that lead to predefined unfolding locations, effectively leading to a weakening of the structure. Presence of the stutter leads to a 'composite' nanostructure that combines a perfect CC with two parallel aligned AH molecules. We note that the weakening only persists in force-extension regimes I and II; the strain hardening regime III is not affected by the presence of the stutter (for a definition of regimes see Figure 5.16).

The stutter only appears in the 2B segment of the vimentin dimer. Notably, this segment is the longest of all segments in the structure (the length of the 2A segment is 158 Å, 139 Å of the 1B segment, 48 Å of the 1A segment and 26 Å of the 2A segment). This corroborates the notion that its significance is related to assist redistribution of plastic strains during unfolding. Furthermore, based on the observed unfolding behavior – namely the occurrence of molecular unfolding at the ends – we believe that the linkers between the CC segments may have a similar role.

The concept of division of larger structures into small, nano-sized segments with material properties of complementary nature is observed almost universally, such as in bone (hard and brittle mineral phase combined with a soft collagen phase) [182], nacre (hard mineral phase combined with a soft polymer glue) [183] spider silk (strong beta-sheet crystals embedded in a soft, amorphous matrix) [184] and many other structural biological materials [15, 185].

Figure 5.19 summarizes the effect of different molecular features. Formation of a CC super helix out of individual AHs leads to a strengthening effect. Addition of the stutter provides a molecular defect that lead to predefined unfolding locations, effectively leading to a slight weakening of the structure while leading to a more homogeneous distribution of plastic strains. Weakening only persists in force-extension regimes I and II (in Figure 5.16); the strain hardening regime III is not affected by the presence of the stutter. Biologically, this makes sense since IF networks are thought to be 'invisible' at small deformation and are only activated at large strains. Thus the stutter helps to realize this trait.

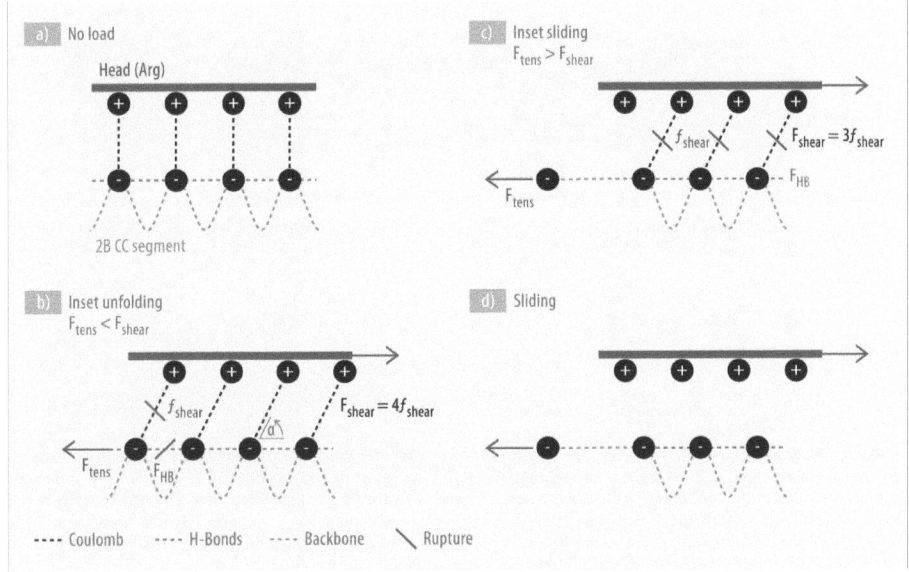

Figure 5.21: Basic deformation mechanisms: In this example $k_c = 3$. As shown in subplot (b) unfolding occurs as long as the number of existing Coulomb bonds is higher than k_c. Once k_c is reached shearing instead of unfolding appears (subplot (c)). The number of k_c depends on the pulling speed. At high speeds, when the molecule is very strong shearing will set in, even if several Coulomb bonds are present. At low deformation rates unfolding will take place until only one Coulomb bond is present, leading to small values of k_c. Subplot (d) depicts pure sliding as deformation mechanism, when all Coulomb bonds are broken.

5.4 Quaternary structures: concurring mechanisms of coiled-coil tetramers

In this Section we present a brief analysis of the deformation mode at the next higher hierarchical scale in the vimentin network, *i.e.* the mechanical behavior of IF tetramers (see Figure 2.3 and 5.20) [186]. This analysis should illustrate how the detailed insight into the fundamental deformation mechanism at the single protein level can help biologists in understanding the deformation mechanisms at higher scales, in this case the interaction between two CC proteins.

5.4.1 Protein structure

Figure 5.20 depicts how interdimer adhesion is facilitated. Extensive experimental analyses of the assembly process revealed that the adhesion between two CCs is dominated by the interaction of the head domain [23, 29, 187]. The attraction between the head and the CC domains of both proteins is mainly electrostatic. Hereby the head domain is primarily positively charged at pH 7 (due to approximately 15% Arg residues), and the CC 2B of the second dimer is primarily negatively charged at pH 7. The head is

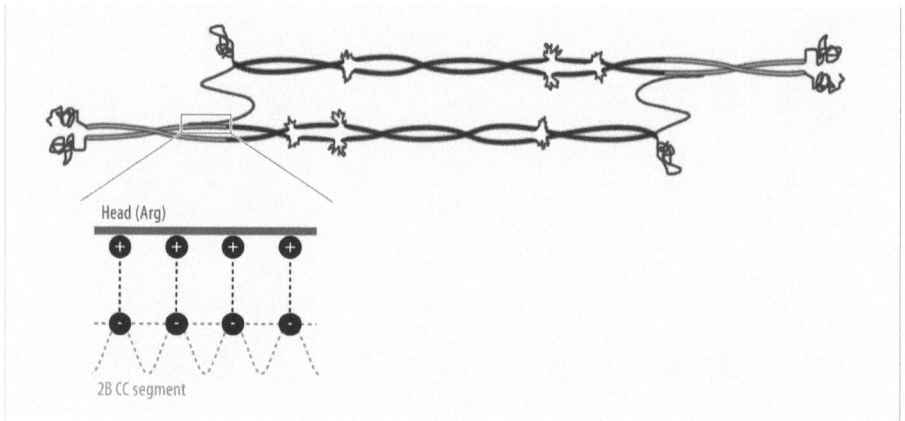

Figure 5.20: Schematic of the tetramer, formed by interaction of two dimers. Two dimers are assembled in the A_{11} assembly pattern antiparallel, approximately half-staggered, as suggested in experimental studies [23]. Interdimer adhesion is facilitated by the interaction of the head domain with the other dimer (indicated with the red color). The head domain is primarily positively charged. It folds back to the negatively charged CC 1A domain and also connects to the 2B domain of the other dimer, which is also negatively charged. The details of the interdimer bonding are schematically shown in the blow-up (lower part of the figure).

simultaneously connected to its own 1A domain, which is negatively charged as well (the 1A and 2B domains contain approximately 11% Glu and Asp residues). Details of the interdimer bonding are schematically shown in the blow-up (lower part of Figure 5.20).

5.4.2 Results of theoretical estimates

Under tensile loading of a vimentin tetramer, the forces are distributed predominantly as tensile load carried by individual CCs, and as shear forces between different CC dimers (see Figure 5.21). The key to arrive at insight into this question is therefore to consider the interplay of two competing mechanisms:

(1) *Molecular unfolding* of the CC dimer, mediated by rupture of HBs (as discussed in previous sections), characterized by $F_{tens}(v) \sim \ln(v)$. Thereby the strength of the dimer depends on the pulling speed, making the molecule very strong at high pulling rates and very weak at low pulling speeds.

(2) *Interdimer sliding*, mediated by rupture of the head-CC Coulomb bonds, characterized by $F_{shear}(k_c) \sim k_c$.

Thereby v is the pulling speed and k_c is the number of Coulomb bonds being involved ranging from n_c to 0 (in our case $n_c = 4$).

The schematics of the different bond-breaking mechanisms are shown in Figure 5.21. Thereby we assume that the head cannot be stretched (random coil), whereas the CC can unfold once the HBs are broken. We note that unfolding in the region of Coulomb interactions will require in addition to the rupture of existing HBs the rupture of at least one Coulomb bond (see subplot 5.21 (b), also expressed in Equation (5.7)).

The interplay of both mechanisms results in three possible failure schemes, where (I) and (II) are the extreme cases and (III) the intermediate (general) case, which is a combination of (I) and (II):

(I) Pure sliding (simultaneous rupture of all Coulomb bonds) under the condition that the sum of all Coulomb interactions is smaller than the strength of the dimer: $F_{shear}(n_c) < F_{tens}(v)$. This is most probable at very high pulling speeds, when the dimer becomes very "stiff" and undeformable.

(II) Complete unfolding of the relevant protein domain under the condition that the Coulomb interaction of individual residues is stronger than the strength of a dimer: $F_{shear}(1) = f_{shear} > F_{tens}(v)$. This is most probable at very low pulling rates, when the dimer is very "soft".

(III) Combination of both schemes, i.e. unfolding of the protein (II) until a critical number of Coulomb bonds (k_c) is reached, followed by sliding (I): The onset of sliding is characterized by $F_{shear}(k_c) < F_{tens}(v)$.

In summary, the relation we are interested in is the number of remaining Coulomb bonds at which sliding will begin in dependence of the deformation velocity, expressed by $k_c = k_c(v)$. This primarily expresses case (III), which is the most general case. In a second step this can easily be linked to the overall extension or strain of the entire tetramer structure.

From the schematic in Figure 5.21 (b) we can see that the force necessary for CC unfolding can be calculated as follows:

$$F_{tens}(v) = \frac{1}{2} \cdot \left(F_{HB} + \frac{f_{shear}}{\cos(\alpha_c)} \right) = \frac{1}{2} \cdot \left(a \ln(v) + b + \frac{f_{shear}}{\cos(\alpha_c)} \right). \tag{5.7}$$

Once the unfolding wave has reached the first Coulomb bond Equation (5.7) becomes the necessary condition for unfolding: In addition to protein unfolding (see Equation (4.4) or (4.18)) the first Coulomb bond must be broken. Thereby the factor 1/2 needs to be taken into account as two parallel dimers are stretched simultaneously (Figure 5.20).

In contrast to that shear will appear under the condition that k_c remaining bonds are broken:

$$F_{shear}(k_c) = k_c \cdot \frac{f_{shear}}{\cos(\alpha_c)}, \tag{5.8}$$

where f_{shear} is the force per bond. As shown above, CC unfolding will be displaced by shearing under the condition: $F_{shear}(k_c) < F_{tens}(v)$. By combining Equations (5.7) and (5.8) we receive the following expression for k_c:

$$k_c(v) < \frac{\cos(\alpha_c)}{2 \cdot f_{shear}} \cdot (a \cdot \ln(v) + b) + \frac{1}{2}. \tag{5.9}$$

This means that the protein will start to shear, when the number of remaining bonds k_c is smaller than the ratio calculated at the right hand side of the equation. In other words, increasing deformation speed will make the dimer stronger (as a is positive) and thus more Coulomb bonds k_c will be necessary to prevent the structure from shearing. The relation between the number of Coulomb bonds and the pulling velocity is shown in Figure 5.22.

For the calculation we assume for simplicity $\alpha_c = 45°$ (see Figure 5.21). Further, $a = 35\,pN$ and $b = 600\,pN$ are estimated by linking experimental results to simulation results on CCs (similar as in the SDM for individual AHs in Figure 5.13 (a)). Shearing forces per bond $f_{shear} = 84\,pN$ were calculated following the Coulomb law [88, 91, 93]:

$$f_{shear} = \frac{1}{4\cdot\pi\cdot\varepsilon_0}\cdot\frac{|q_1|\cdot|q_2|}{r_0}, \quad (5.10)$$

where $\varepsilon_0 =$ 8.85E-12 C^2 N^{-1} m^{-2} (permittivity of free space), and $|q_i|$ describes the two charges (modeled as point charges). Note that r_0 refers to the spacing of the particular Coulomb bonds considered. We take for $r_0 \approx 5\,\text{Å}$, following the results from recent experiments, where the space between two dimers was measured in the order of 15 Å [187]. We divide this distance by 3, in order to cover (i) the distance between the 1A-segment and the head, (ii) the distance between the head domains and (iii) the distance between the head and the 2B-segment (see Figure 5.20). It is assumed that the interdimer bonds are formed between the side chains of the participating AA residues without presence of water molecules between the bonds.

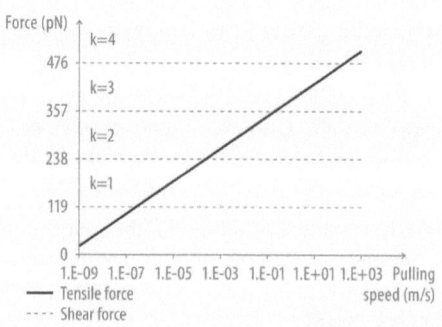

Figure 5.22: shows the relation between the pulling speed and the critical number of remaining bonds at which shearing will set in. At high pulling speeds shearing will set in, even if several coulomb bonds are present (e.g. between 0.1m/s and 15m/s when k=3 bonds remain), whereas at low pulling speeds unfolding will take place until only one coulomb bond remains. At pulling speeds smaller than 1E-7 m/s not shearing will take place, as at that pulling sped coulomb bonds are too strong in order to be broken.

Parts of the following derivations concerning the partial charges $|q_i|$ were already published earlier [188] and were adopted for this analysis. The electric charges $|q_i|$ can be determined from an analysis of the atomistic charge distribution of the individual residues. All considerations are carried out at pH 7. The head domain contains positive charges due to Arginine (Arg) residues. As the Arg side chain has a pK$_a$=12.48, it is charged positively at pH 7. Following the charge distribution predicted by the CHARMM force field [91], the resulting partial charge of the end of the side chain is $|q_1|$= +0.12 elementary charges. The 1A/2B segment of the CC domain contains strongly negative charges due to presence of glutamic acid (Glu) residues with pK$_a$ =4.07, as well as aspartic acid (Asp) residues with pK$_a$ =3.86. In Asp and Glu, the side chain is consequently protonated, with a partial charge of $-|q_2|$= -0.76 elementary charge at the end of the side chain [91].

We estimate the number of possible bonds that can be formed between the head and CC domain by considering the details of the head-CC interactions. Experiments suggest that each Coulomb bond is build out of an interaction of an Arg residue in the head domain, with either an Asp or a Glu residue in the CC segment. The analysis of the AA sequence in the head domain and CC domain reveals that about 11-14% of residues are charged residues [54]. Consequently, there are approximately 4 Arg residues in contact with the CC. We calculate this number by assuming that half of the 80 residues in the head fold onto the dimer (see Figure 5.20) and keep in mind that about 11-14% of residues are polar. In consequence, approximately $n_c = 4$ interdimer 'bonds' can form in the overlap region between the head domain and the 2B segment of the CC.

5.4.3 Conclusions in light of materials science and biological function

As reported in Section 2, vimentin IF play a crucial role in determining the large deformation behavior of eukaryotic cells. In this Section we have presented a simple analysis of the interdimer adhesion of vimentin IFs. Our analysis provides first insight into adhesion forces that lead to CC dimer unfolding or interdimer shearing. Such models are necessary to advance the understanding of biological processes like mechanotransduction, which critically depend on the nanomechanical properties of IFs.

The simple analysis put forward here suggests that the interdimer adhesion provided by the particular interaction of Arg residues with Glu and Asp residues may be at the borderline between CC unfolding at low pulling rates and interdimer sliding at high pulling rates. This suggests a possible balance between both mechanisms, increasing the strength of the structure, when exposed to shock loads while keeping it soft and deformable at vanishing pulling rates so that structural rearrangements can be fulfilled with minimal energy need. This was already shown for individual proteins (Section 5.2). In this example we show that this balance is also valid on the next higher scale the inter-protein scale. If this were indeed the case, the system would be structured so that the optimal shear force is transduced to reach the limiting tensile strength within each CC. In other words, the structure may be adapted to make optimal use of the interdimer 'glue' material.

We emphasize that while this conclusion is very rough, it agrees well with earlier observations in the structural analysis of the collagen fibril structure [47]. It was found that the critical length scale of tropocollagen molecules may be controlled by the driving force to maximize the tensile forces in each molecule and the maximum shear that can be transmitted between molecules. Interestingly, during shearing the amount of dissipated energy increases not only because long ranging Coulomb bonds are broken and rebuild but also due to the amount of dissipation energy arising from water perturbations surrounding the protein. It was already shown for bone that the combination of tensile deformation and shearing is a successful path in creating strong yet tough and stiff materials [45].

The study reported here has several limitations. For instance, the quantification of the interdimer adhesion was achieved using a simplistic model based on Coulomb interactions. Atomistic details of the interdimer bonding remain elusive, and could be addressed by extensive MD simulations with reactive force fields, for instance. Further, the details of the interdimer bond rupture processes are entirely unknown and were calculated based on simple assumptions. Possible entropic effects of the dimer and the head domain were not covered [14]. These effects could, however, significantly influence whether or not interdimer sliding or CC unfolding is the domination mechanism of deformation.

The discussed limitation regarding the structure and the appearing interactions are indicating the current frontier of research, making the intermolecular interaction neither reachable by experiments nor by simulations. In order to gain additional insights new experimental techniques as well as simulation approaches will be necessary. Examples of new efforts were reported recently in [65], where individual molecules were manipulated with a AFM tip perpendicular instead along the protein axis, mimicking a bending test and thus shearing individual proteins against each other.

6 Mechanics of multi-hierarchical systems and protein networks

6.1 Multi-hierarchical systems

6.1.1 Analyzing the behavior of model systems with different hierarchies

In this Section, we show in three examples how the theory derived in Section 4 can be applied in order to better understand and further synthesise HBMs. We note that these are model cases, and not each system studied here is found *in vivo*. However, these analyses are important to develop a detailed understanding for future materials synthesis approaches.

First we present an application of our theoretical model in an analysis of strength and robustness properties of AH based protein structures. The main question of the following

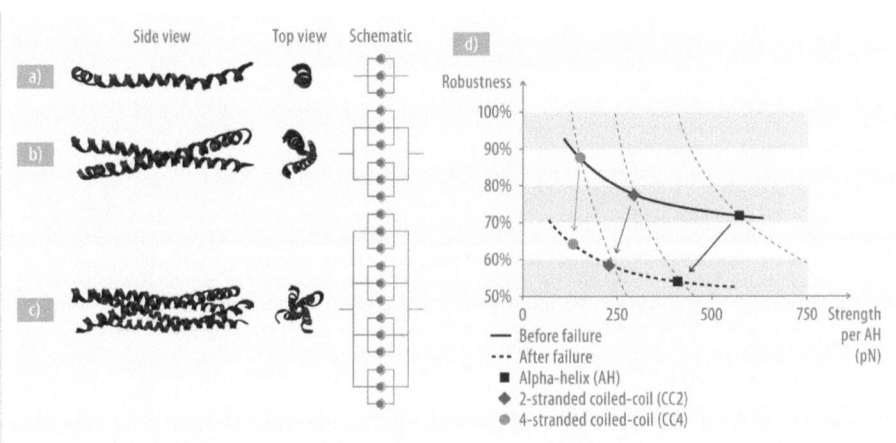

Figure 6.1: Overview over three model systems, including (a) single AH, (b) CC with two AHs (CC2) and (c) a CC protein with four AHs (CC4). These system were already used for theory validation in Section 4.5.1. The schematic on the right shows the abstract representation of each physical system. Each blue line represents a single HB. In the atomistic simulations, all structures are embedded entirely in a water skin during deformation; however, water molecules are not shown here for clarity. Subplot (d): Performance of the analyzed structures in the strength per AH-robustness space, before and after failure. The coarsely dashed lines represent levels of equal strength (s) - robustness (r) potential (that is, the product of both values is equal on these lines, r·s = const.). Robustness and strength compete on these lines. The figure illustrates that vimentin CCs (CC2) have a robustness degree of 80%. The first data point for each structure represents the intact system, whereas the second data point shows the system after failure. Thereby, robustness is defined as parameter insensitivity following Equation (5.5). In this case robustness equals to the force from hierarchical strengthening of a defect system (two instead of three HBs rupture simultaneously) divided by the force of an intact system (all three HBs rupture at once). As we can see, a defect in a system moves the system to another potential line. For example, due to the high level of robustness, the CC4 structure hardly changes its strength, whereas the strength of a single AH is significantly reduced. This illustrates how different combinations of robustness and strength can be combined through structural design.

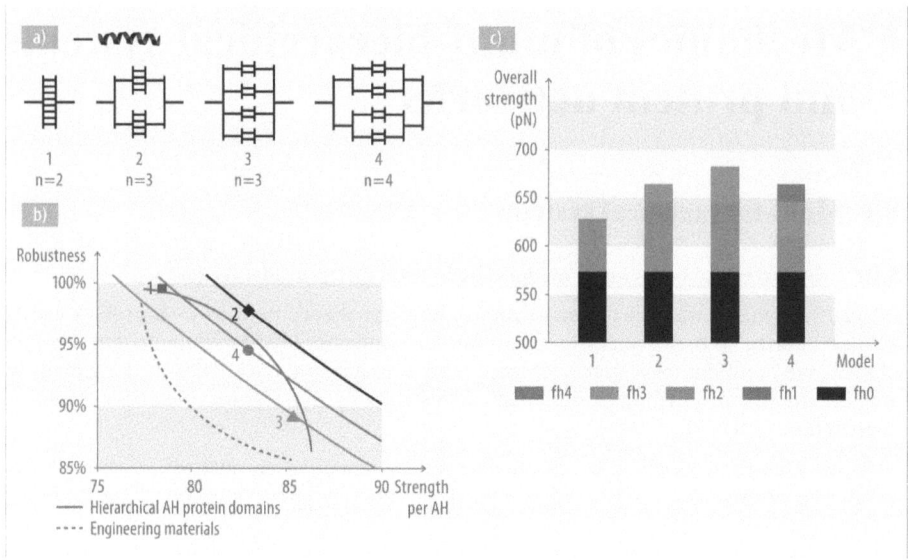

Figure 6.2: This figure shows an example of four different structures with the same number of subelements (that is, eight AHs) but in different hierarchical arrangements. Subplot (a) shows the four different architectures. For simplicity, individual HBs on the lowest hierarchical scale are not shown; instead one line represents three HBs as one AH. The number n describes the number of hierarchies present in this system. Subplot (b) shows the concurrence between strength and robustness, which depends on the degree of redundancies on different hierarchical levels. The level of robustness increases with increasing redundancies on a particular level. Dependent on the hierarchical arrangement of the elements, different potentials of strength and robustness can be reached. Thereby a reinforcing combination of robustness and strength is possible. In contrast to that in engineering materials one dimension is maximized to the cost of the other, as indicated here with the "banana curve". Subplot (c) shows the contributions of each hierarchy to the overall strength (not strength per AH, as shown in subplot (b)). As we assume that in each AH three HBs break simultaneously, each structure (featuring an AH as the smallest subelement) has the same contribution from hierarchy 0. This is also the highest amount of strength contribution and shows the significance of the strength of HBs, which depends on the solvent and the environment. The other contributions are of "hierarchical" origin. The force contribution from hierarchy 1 is zero, since 3 out of 3 HBs break, which lowers the logarithmic multiplier to zero.

discussion is: How are protein materials capable of unifying strength and robustness. Thereby no length scale effects, as shown in Equation (4.24) are considered. We calculate the robustness following Equation (5.5). After discussion the structure already introduced in Section 4.5 for theory validation, we present two theoretical analyses: The first one illustrates the effect of increasing the number of elements at the same hierarchical scales. The second analysis illustrates, how introduction of multiple hierarchies affects the performance of the system.

We begin with an analysis of three AH-based structures (same structures as discussed in the validation in section 4.5.1). Figure 6.1 (a)-(c) depicts the architecture of the considered structures, and Figure 6.1 (d) shows the performance of the analyzed structures in the strength per AH-robustness space before and after failure of one basic element. It is apparent that strength and robustness are not completely independent parameters; with increasing number of hierarchies, the robustness increases due to the increasing number of redundancies present in the structure. However, this goes hand in hand with a loss of strength per element (that is, per AH), since the overall strength of a structure is not directly proportional to the number of parallel elements, but logarithmically as shown in Equation (4.25). The strength of the failed system would be highest for an infinite number of elements on each scale – leading to a robustness value of 100%. However, increasing the number of elements at a specific hierarchical scale is inefficient, as it leads to extensive material use and a decrease in strength (see e.g. Figure 5.15).

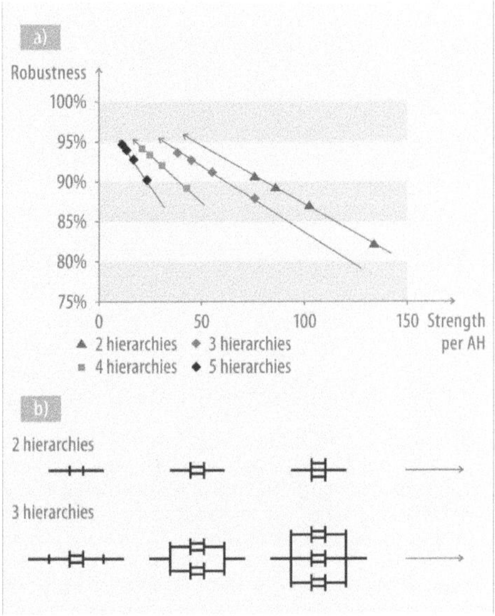

Therefore, the introduction of multiple hierarchies becomes significant. To illustrate this, we arrange eight single AHs in different hierarchical structures – asking the question: How can one arrange eight AHs to obtain various levels of robustness and strength? As shown in Figure 6.2, the systems consist of two, three and four hierarchies. The differences in robustness and strength are remarkable, since they are not achieved through additional use of materials, but purely through different hierarchical arrangements. To the best of our knowledge, this tuning of properties in the strength-robustness domain as illustrated here has been shown for the first time. In the robustness-strength map, the 'best' material behavior is the one in which high robustness is achieved at large strength – referred to as a 'high potential' in the following discussion. It can be seen that, remarkably, system 2 has the highest potential. It is notably not the system with the highest hierarchical level (system 4), nor the system with the highest level of redundancies (system 1). In other words, system 2 features the best combination of redundancies on the different levels.

Figure 6.3: Robustness as a function of strength per AH. The direction of the arrow goes into the increasing number of sub-elements on the particular hierarchy. Subplot (b) depicts examples for 2-hierachy and 3-hierarchy systems, each with one, two and three elements. These results show that in general, the more elements, the more robust but the less strong (per AH) is a system. This also shows the tradeoff between robustness and strength. We use strength per AH due to a better comparison of different structures, which have different amounts of material (AHs) per cross-sectional area. This equals to a normalization of force by the cross-sectional area, which leads to the strength of a material (similar to stress, which equals to force normalized by the surface area).

As a third example that illustrates an application of the theory derived in Section 4, Figure 6.3 depicts a systematic investigation of the robustness-strength behavior under an increasing number of elements at each hierarchy. Each line represents one level of hierarchies (e.g. 2 hierarchies equal to the AH level, 3 hierarchies equal to the CC level), where the number of elements on this particular hierarchies is varied (e.g. one AH, two AHs, for 2 hierarchies system and one CC, two CCs, etc. for a 3 hierarchies system). Consequently, each hierarchy is represented by another equi-hierarchical line. We find

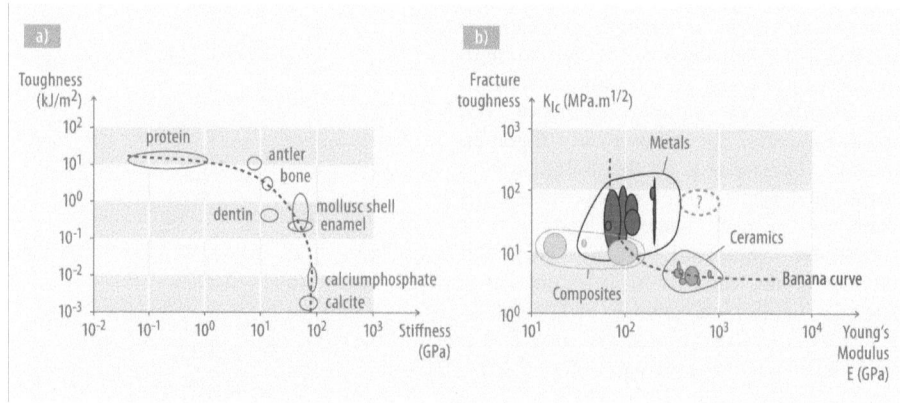

Figure 6.4. Subplot (a) compares the toughness and stiffness properties for a number of biological materials (adapted from Fratzl *et al.* [19]). Biological composites, such as antler, dentin, bone and enamel are result of a "reinforcing" combination of the protein toughness and the mineral-stiffness. Subplot (b) depicts toughness and stiffness values for synthetic materials, such as metals, alloys and ceramics (adapted from Ashby *et al.* [30]). Remarkably, all these materials lie – in contrast to biological materials – on the 'banana-curve', an inverse relation between increasing toughness and decreasing stiffness. The yellow region shows the property region of high toughness and stiffness that may be accessible through designing bio-inspired materials.

that on each "equi-hierarchical" line with an increasing number of elements the strength per AH decreases, while the robustness increases. This example also shows that with different arrangements of equal elements, almost any point in the strength-robustness space can be achieved, allowing one to overcome the competition between strength and robustness purely by structural rearrangement, without inventing new elements.

6.1.2 Conclusions in light of materials science and biological function

With different structural arrangements, different combinations of strength and robustness can be achieved [189]. This finding is the most important result of these examples: It illustrates that the conflict between strength and robustness can be resolved by introducing hierarchies as an additional design variable. These results further suggest that the level of hierarchical depth and strength may be balanced in biological protein materials, since the robustness and strength are not completely inversely proportional, allowing biological materials to maximize the mechanical performance while minimizing the use of materials. Overall our analysis illustrates that the introduction of hierarchies is the key to unify disparate properties. Applying this insight will allow an extended use of hierarchies in bio-inspired or biomimetic synthetic materials at nanoscale, such as hierarchically organized CNT-bundles, nanowires or polymer-protein composites[190-192].

Further, the examples analyzed above show that different mechanical properties can be achieved purely through structural rearrangement instead of inventing new building blocks. Broad application of universal building blocks in highly diverse architectures might be a strategy Nature follows consequently, where the specific structure evolves in dependence of the environmental requirements. Adaptation to changes in the environment can be achieved simply by adopting the structure. Applying this concept to synthetic materials might result in smart structures, which continuously and independently adapt to environmental changes, making the invention of new elementary building blocks unnecessary. More detailed discussion on this topic will follow in Section 7.3.

Broadening the perspective: strength-toughness maps

The search for materials that combine contrasting properties, such as high strength and high toughness, has driven the design of many alloys and composites. However, most synthetic materials fail to achieve the combination of disparate properties (and thus lie on the so-called "banana curve" in regard to these properties), consequently possessing one at the cost of the other (Figure 6.4 (b)) [45]. Biological materials, e.g. collagenous materials such as bone [193], nacre [194, 195] or dentin [196], on the other hand, are stiff and tough (Figure 6.4 (a)). Materials composed of protein and mineral phases are thus almost as tough as the protein phase and as stiff as the mineral phase.

In the style of stiffness and toughness analysis, we have shown above that strength and robustness can be unified through structural modifications in biological materials at nanoscale. Here we describe shortly, how these properties are linked to each other: The strength of a material is determined on the atomistic level by the height of the energy barrier of individual bonds and is known as the yield strength at the end of the elastic deformation regime in a stress-strain curve (for example, seen as the AP in the deformation of AHs). The stiffness of a material is proportional to the curvature of a bond's energy well and thus determines the slope of the elastic regime in the stress-strain curve (known as the Young's modulus). Most materials have a strong correlation between strength and stiffness due to an upper limit of bond interactions (represented by the width of energy well). This restriction leads to a higher curvature with increasing well depth, *i.e.* to higher stiffness with increasing strength [197].

Similarly, robustness and toughness are linked to each other. Toughness is described by the amount of dissipated energy a material can withstand before it breaks, mainly determined by the ability to withstand large degrees of plastic deformation before catastrophic failure. This behavior is referred to as fault tolerance (e.g. mediated through the existence of dislocations or small cracks in ductile metals that prevent such catastrophic failure). Materials with small degree of toughness (often referred to as "brittle") break already when small defects are present and thus are not very robust.

Here we have focused on an analysis of strength and robustness, since these are key engineering properties that determine the fracture behavior of materials either in biological or engineering applications. Further, robustness as system property is more common in system biology than toughness, which commonly regarded as a material property.

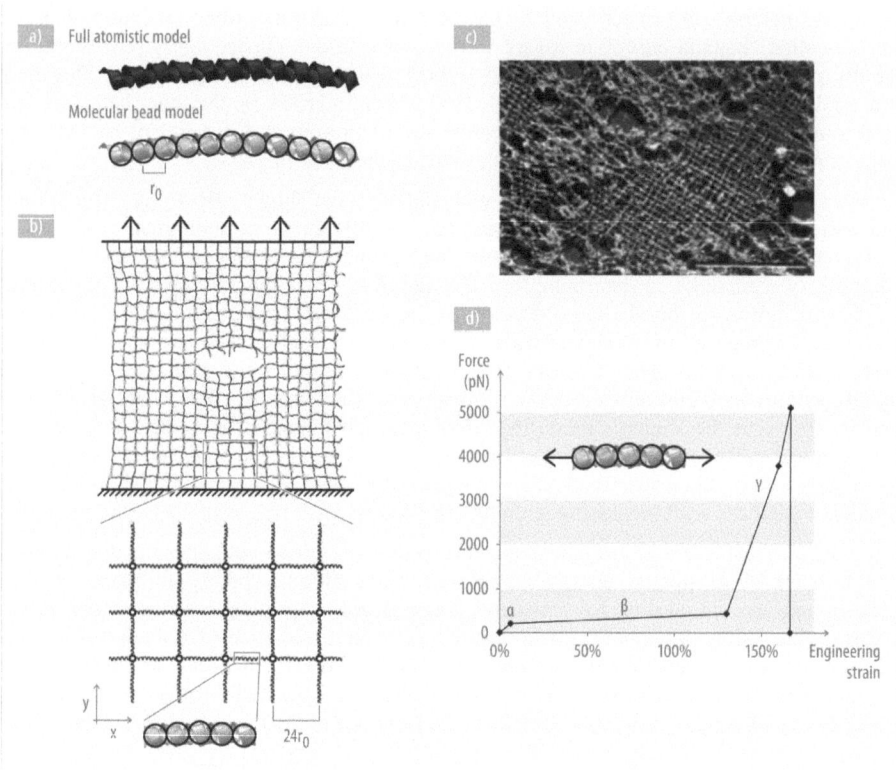

Figure 6.5: Subplot (a) shows a schematic of the coarse-graining procedure, replacing a full atomistic representation of an alpha helical protein domain by a mesoscale bead model with bead distance r_0. Subplot (b) depicts a schematic of the coarse grained protein network used in this study with the applied boundary conditions. The size of the network equals to 240×240 nm. Thereby each filament is represented by one AH as shown in the blow-up. A constant strain rate is applied in y-direction. We study networks with and without cracks (here an example with crack; the crack represents geometrical flaws/inhomogeneities as they appear *in vivo*). Subplot (c) shows a snapshot of a quasi-regular lamin meshwork as it was observed in experiment including geometrical irregularities (scale bar 5 μm). Figure reprinted from Aebi et al. [28]. Subplot (d) depicts characteristic force-strain curves for pulling individual alpha-helices as used in our mesoscale bead model. As explained in Materials and Methods, this force-strain behavior is derived from full-atomistic simulations and theoretical analysis, and has been validated against experimental studies. The labels C, C and C identify the three major regimes of deformation (see Figure 5.13). We assume covalent bond rupture of the backbone at forces of 7,800 pN, as calculated with ReaxFF reactive force fields in previous studies (not shown here) [38].

6.2 Flaw tolerance of alpha-helical protein networks

The focus of this Chapter is on understanding the role of the alpha-helical protein motif under mechanical deformation of larger-scale protein networks, without and with structural (geometric) imperfections. To achieve this, we consider the deformation and rupture behavior of a simple model of an alpha-helix based protein network, as shown in Figure 6.5 (b). We emphasize that the goal of our model is not to accurately reflect a particular type of a protein structure. Rather, it is formulated deliberately as a general model to probe fundamental properties of a broader class of protein materials in which alpha-helix based protein filaments connect to form larger-scale networks. Despite its simplicity, our mesoscale simulations as introduced in Section 3.3 captures the essential physical properties of individual alpha-helical protein filaments as discussed in previous sections.

Through simulation of a larger-scale network, our model enables us to provide an important link between single molecule properties and mechanisms and the overall material behavior at much larger length- and time scalescales. Scales in the order of magnitude of microseconds and micrometers are indispensable for the design of novel research studies, as many critical biological processes such as mechanosensation or mechanotransduction in cells appear at this scale. Mesoscale models feature fewer degrees of freedom, but are still capable of capturing the essential physical properties of the molecules. Here we demonstrate how such a model can be developed and applied for studies on protein networks consisting of AHs, as they were discussed in Section 5.2. Whereas in MD simulations we did not overcome a length scale of 7 nm and a timescale of tens of ns, here we perform a simple study on a length scale of 0.12 μm and timescale of 0.2 μs, with a great potential toward simulation the dynamic behavior of even larger systems.

In the literature, alpha-helical protein materials have been studied either from a macroscale perspective or from a single-molecule level, but not from an intermediate "mesoscale" viewpoint. For example, alpha-helix based intermediate filament networks have been investigated through shear experiments of protein gels [42] as well as through in situ studies with particle tracking rheology [198], where their material properties have been explored from a macroscopic perspective. On the other hand, the mechanical properties of the elementary nanoscale alpha-helical building blocks were studied extensively, and several publications have reported advances in the understanding of their nanomechanical behavior from both experimental [167, 168] and theoretical [14, 24, 26, 199, 200] perspectives.

Up until now the properties of alpha-helical protein networks specifically at the mesoscale have not yet been investigated, and no analysis of the rupture behavior of these networks was reported, despite their widely accepted significance of the mechanical performance and integrity. This has thus far hindered the formulation of bottom-up models that describe the structure-property relationships in protein networks under large deformation, which may explain their characteristic mechanical behavior. In particular, it remains unknown what the mechanism is by which these protein networks can sustain such large deformation of several hundred percent without catastrophic failure. This is an intriguing question since protein networks typically feature structural irregularities and flaws in their network makeup, as highlighted in Figure 6.5 (c). In synthetic materials (such as polymers, metals or ceramics), flaws typically lead to catastrophic failure at relatively small strains (often less than a few percent), preventing a material from undergoing very large deformation, reliably. This is because crack-like imperfections are generally responsible for initiating catastrophic failure [197], because they lead to very large stress concentrations at the corner of the cracks.

6.2.1 Modeling approach

Model formulation

The basis for the network model is a coarse-grained description of an alpha-helical protein structure, referred to as a mesoscale bead model. In our model, the entire sequence of amino acids that makes up the alpha-helix structure is replaced by a collection of beads (see schematic in Figure 6.5 (a), where each bead represents hundreds of atoms in explicit solvent. This approach is adapted since it significantly reduces the computational cost of simulating a large protein network, enabling us to describe a large lattice-like network of strongly bonded alpha-helices (Figure 6.5. (c)) (these bonds may be formed through intermolecular cross-links or strong electrostatic bonding). The beads in the mesoscale model interact according to an intermolecular multi-body potential, developed to reflect the key physical properties of individual alpha-helical protein domains including adhesion, stretching and bending. The total energy of the system is given by

$$U(\vec{R}) = U_T + U_B,$$ (6.1)

where $\vec{R}$ denotes the positions of all particles. The total energy is given by the sum over all pair-wise (that is, U_T) and all three-body interactions (that is, U_B), where

$$U_T = \sum_{pairs} \phi_T(r) \quad \text{and} \quad U_B = \sum_{angles} \phi_B(\varphi)$$ (6.2)

Specific interparticle potential energy expressions are defined for each of the contributions. We approximate the nonlinear force-extension behavior of alpha-helical proteins under tension by a multi-linear model. This multi-linear model is a combination of four spring constants $k_T^{(i)}$ (i = 1..4), which are turned on at specific values of molecular stretch. A similar model has been used successfully in earlier studies of fracture in crystalline model materials [83, 201] and provides an effective way to describe the nonlinear constitutive behavior based on computationally effective, simple piecewise harmonic potential functions. Based on this model, the tensile force between two bead particles is described as:

$$F_T(r) = -\partial \phi_T(r)/\partial r,$$ (6.3)

(the energy function U_T is given by integrating the force $F_T(r)$ over the radial distance), where

$$\frac{\partial \phi_T}{\partial r}(r) = H(r_{break} - r) \begin{cases} k_T^{(1)}(r - r_0) & r_1 > r \\ R_1 + k_T^{(2)}(r - r_1) & r_1 \leq r < r_2 \\ R_2 + R_1 + k_T^{(3)}(r - r_2) & r_2 \leq r < r_3 \\ R_3 + R_2 + R_1 + k_T^{(4)}(r - r_3) & r_3 \leq r \end{cases}.$$ (6.4)

In eq. (6.4), $H(r - r_{break})$ is the Heaviside function $H(a)$, which is defined to be zero for $a < 0$, and one for $a \geq 0$. The parameters $R_1 = k_T^{(1)}(r_1 - r_0)$, $R_2 = k_T^{(2)}(r_2 - r_1)$ and $R_3 = k_T^{(3)}(r_3 - r_2)$ are calculated from force continuity conditions. The bending energy of a triplet of three bead particles is given by

$$\phi_B(\varphi) = \frac{1}{2} K_B (\varphi - \varphi_0)^2 \qquad (6.5)$$

with K_B relating to the bending stiffness of the molecule EI through $K_B = 3/2 EI / r_0$.

Model parameter identification

All parameters in the mesoscale bead model are determined from full atomistic simulation results and theoretical studies, based on careful studies reported in Section 5.2 and earlier publications that involve experimental validation [24, 26, 38]. We choose $r_0 = 0.5$ nm per bead, providing significant computational speedup while maintaining a sufficiently fine discretization of the alpha-helical protein (leading to a bead particle mass m = 400 amu). All parameters in eq. (6.4) are fitted to reproduce the nanomechanical behavior obtained using the full atomistic model with the molecular formulation (Section 5.2). In particular, the stiffness in regime α (in Figure 6.5 (d)) $k_T^{(1)}$ is identified from these simulations. Further, a detailed analysis of the alpha helix behavior in dependence of the deformation rate was carried out in previous studies, where it was shown that for vanishing pulling rates the force at end of the first regime (see regime α) and the beginning of the second regime (see regime β) reaches an asymptotic value of ≈200 pN (Section 5.2, [200]). It was also shown that this value agrees with experimental measurements and we thus consider this in the formulation of our bead model to mimic quasi static deformation at vanishing pulling rates as relevant for physiological and experimental deformation speeds. This enables us to identify the onset point for the second regime, r_1. The stiffness in regime β $k_T^{(2)}$ is identified from atomistic simulations [26]. The onset of regime β, described by parameter r_2. is identified from atomistic simulation as well [38], which includes specifically the extraction of the transition strain and the stiffness parameters in regime γ (that is, $k_T^{(3)}$ and $k_T^{(4)}$ as well as r_3).

Bond rupture of the protein polypeptide backbone is modeled at forces of ≈5,500 pN, which provides the value for r_{break}. This is based on earlier ReaxFF reactive force field results [38] (here we use a slightly smaller value for the rupture force than reported in [38] to reflect the behavior at vanishing pulling rates). The bond strength of several nN for strong bonds as used here is a value widely accepted in the literature

Equilibrium bead distance (in Å)	r_0	5.00
Critical distances (in Å)	r_1	5.90
	r_2	11.50
	r_3	13.00
Tensile stiffness parameters (in kcal/ mol/ Å²)	k^1_T	9.70
	k^2_T	0.56
	k^3_T	32.20
	k^4_T	54.60
Bond breaking distance r_{break} (in Å)		13.35
Equilibrium angle φ_0 (in degrees)		180.00
Bending stiffness parameter k_B (in kcal/ mol/ rad²)		3.44
Mass of each mesoscale particle (in amu)		400.00

Table 6.1: Summary of the parameters used in the mesoscopic molecular model, chosen based on full atomistic modeling of AH molecules (note that 1 kcal/mol/Å = 69.479 pN).

and has also been measured experimentally [202]. Figure 6.5 (d) depicts the force-strain curve for alpha-helices as reproduced by the mesoscale bead model. The bending stiffness is obtained from bending deformation calculations of alpha-helical molecules, as described in earlier publications [24, 38] (values are validated by comparison with the experimentally measured persistence length on the order of a few nanometers). The time step is chosen to be 15 fs. The entire set of parameters of the mesoscale model is summarized in Table 6.1.

System definition, geometry and boundary conditions

We create a network with a mesh side length of 12 nm (in square shape), which equals to 24 beads since $r_0 = 0.5$ nm per bead (see Figure 6.5 (a)-(b)). The linkers between the perpendicular filaments are modeled as beads that are freely deformable in both directions without any angular restraints. This mimics the existence of cross-links between individual alpha-helical filaments (facilitated e.g. through side-chain mediated bonds, such as disulfide bonding). As shown in Figure 6.5. (b), we create a square meshed protein network out of individual filaments, where each filament consists of a single alpha helix. The orthogonal arrangement of protein filaments roughly mimics an intermediate filament protein network as for example observed in lamins in the nuclear membrane of oocytes (see Figure 6.5 (c)) [28].

We note that the choice of a single alpha helix per filament represents a limitation compared with the actual structure of intermediate filaments in vivo, which typically contains multiple alpha-helices arranged in parallel. However, the purpose of the present study is not to exactly model the structure of lamin, but rather provide a generic study on the behavior of alpha helical protein networks without and with defects. We deliberately avoid the attempt to model a specific protein filament.

We consider a system with 20 filaments (each composed of 24 beads as discussed above); with an overall network size of 240 nm ×240 nm. Pulling is applied in y-direction in mode I tensile loading, as indicated in Figure 6.5 (b). Thereby the first two rows of beads at the bottom are fixed. Displacement boundary conditions are applied to the upper three rows of beads, so that the upper three rows of beads are moved continuously following a prescribed strain rate. A strain rate of $\dot{\varepsilon} = 4.17 \times 10^6$ s-1 is used for all studies (studies with varying strain rates were carried out and it was confirmed that the system undergoes deformation near equilibrium at the strain rate chosen). All simulations are carried out at 300 K in a NVT ensemble (constant temperature, constant volume, and constant number of particles).

The overall length scales reached in this study shows a sufficiently high level of repetition of individual meshes so that boundary effects can be neglected. Larger systems do not change the overall behavior described in this Thesis.

Crack modeling

To model the crack-like inclusion, protein filaments across the crack surface are not connected from the beginning of the simulation (and can not reform). This approach effectively models the existence of a structural imperfection in the protein network through the existence of an elliptical flaw. By controlling how many protein filaments are broken at the beginning of the simulation we control the size of the crack.

Stress and strain calculation

For calculation of stress the virial stress approach was applied [102]. The failure stress is measured at the point when filaments begin to fail (usually identified through a rapid drop of the stress). The failure stress data shown in Figure 6.6 is obtained through an average over the entire simulation domain. The stress at failure shown in Figure 6.7 (b) is defined as the remotely applied stress; which is different than the measured average stress shown in Figure 6.6. It is calculated by taking the applied strain (due to a particular prescribed displacement) and computing the associated stress following the stress-strain response of a perfect crack-free system (i.e. it shows the rupture stress of a system with size $L-\xi$). The strain is defined by $\varepsilon = \Delta L_y / L_y$ (=engineering strain), where ΔL_y is the applied displacement and L_y is the length of the system in the y-direction (the pulling direction).

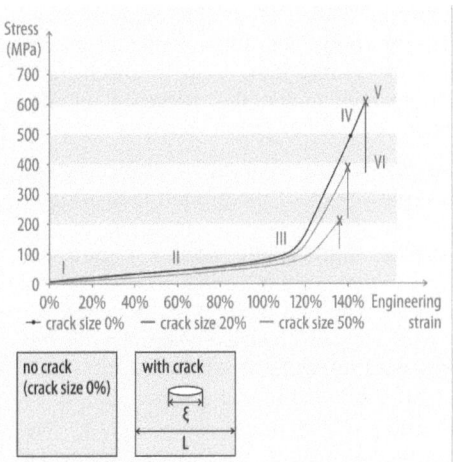

Figure 6.6: Mechanical response of the alpha-helical protein network. The graph shows stress-strain curves of a protein network, with and without a crack, as well as for two different crack sizes. The relative crack size is given as ratio of crack length ξ divided by the system size L, defined as $\chi = \xi/L$. We observe two major regimes, (I-II) a very flat increase in stress until approximately 100 MPa, followed (II-III) by a very steep increase in stress due to strain hardening of the protein backbone up to strains of close to 140..150% (IV). Eventually, (V) covalent bonds of the system break and the entire system fails catastrophically (VI). Interestingly, there exists only little difference in terms of the failure strain between all three systems, indicating the fault tolerance of the studied structure. The perfect system (without a crack) has a strength of ≈600 MPa.

Simulation implementation

The mesoscale simulations following an MD scheme are implemented in the simulation package LAMMPS (Large-scale Atomic/Molecular Massively Parallel Simulator), a MD code allowing simulating short range interactions [203] as introduced above. For visualization we use the OpenDX package. The simulation model implementation in LAMMPS used for our studies is available upon request. All simulations have been carried out at MIT's Laboratory for Atomistic and Molecular Mechanics on a Dell linux computing cluster with Intel Xeon dual core CPUs. One simulation takes approximately 24 hours to complete.

6.2.2 Deformation and rupture of a AH based network

We begin our analysis with carrying a tensile deformation test of an alpha-helical protein network, by using the geometry and loading condition as shown in Figure 6.5 (b). We consider two geometric arrangements. First, a perfect protein network without a structural flaw. Second, a protein network with a structural flaw, here modeled as a crack-like inclusion. The goal of this analysis is to identify how an alpha-helical protein network responds to mechanical deformation under the presence of the crack.

We stretch both systems by displacing the outermost rows of the protein network and measure the stress-strain response of this material, until failure occurs. Figure 6.6 depicts stress-strain curves of the protein network with and without a crack, and for two different crack sizes (where the relative crck size is defined as ration of crack length ξ divided by the systems size in x-direction L, defined as $\chi = \xi/L$). We consider a case $\chi = 20\%$ (the length of the crack is 20% of the size of the structure in the x-direction) and $\chi = 50\%$ (the crack reaches half way through the structure). The purpose of considering different crack sizes is to measure the effect of the size of the structural imperfection on the mechanical properties.

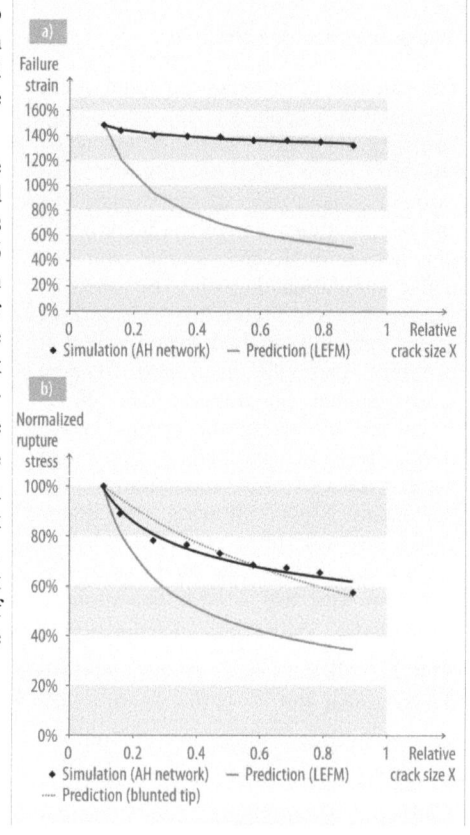

Figure 6.7: Failure strain and failure stress as a function of crack size and comparison with theoretical model. Subplot (a): Systematic analysis of the failure strain of the system, showing the failure strain over the relative crack size χ. The simulation results show that the failure strain is largely insensitive to the presence and size of cracks. Further, the plot includes the prediction based on eq. (6.7), corresponding to a scaling as $\varepsilon_{0,f} \sim \sqrt{1/\chi}$. This behavior reflects that the scaling parameters are much different (-0.0362 vs. -0.5), and that linear elastic fracture mechanics (LEFM) fails to describe the fracture behavior of this material. Subplot (b): Analysis of the failure stress of the system as function of χ. The analysis also shows a deviation from the prediction of LEFM. The blunted crack-tip model is also shown for comparison (dashed line), providing an overall better fit than LEFM through the scaling law $\sigma_{0,f} \sim 1/1+C\chi$. Note that for relative crack sizes < 5% the maximum strain and stress does not change as the material has reached a complete insensitivity with respect to imperfections.

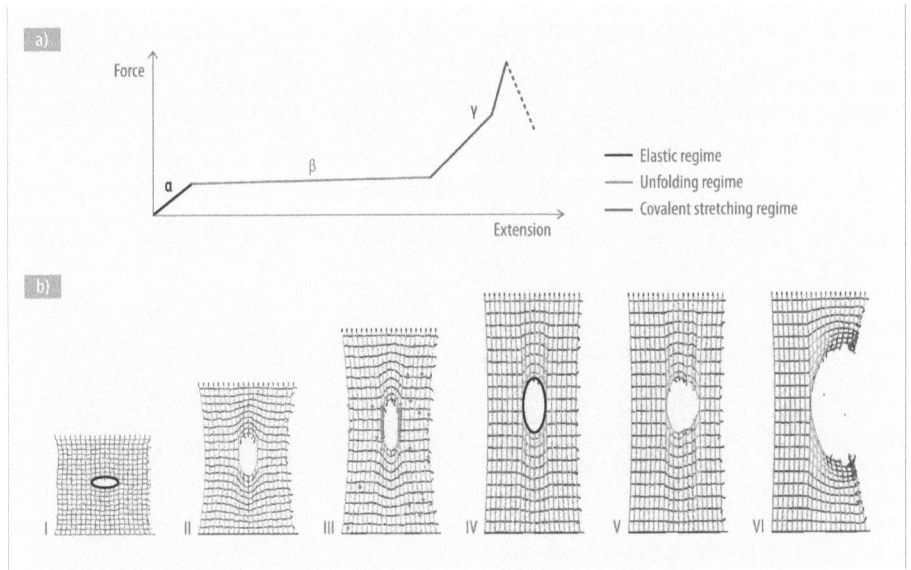

Figure 6.8: Subplot (a) shows a schematic of the characteristic force-extension curve of a single AH (consisting of three regimes) used as input for our simulations. Subplot (b) shows snapshots of the network with crack at different strains. Most interestingly, due to the very large plastic zone no strain localization (colored in red) appears at the crack tip, making the network overall fault tolerant. Only once the whole structure reaches the rupture strain the crack propagates, leading to catastrophic failure. The ellipsoids in Snapshot I and IV highlight the crack shape transformation that occurs during deformation (they show the surface geometry of the crack). The color scheme represents the molecular deformation regime as shown in subplot (a).

For all cases considered we observe two major regimes in the stress-strain response, (I-II) a very flat increase in stress until approximately 100 MPa, followed (II-III) by an increasingly steep increase of the stress, which lasts up to stresses close to 600 MPa (III-IV). Eventually (IV) the strong bonds between different alpha-helical protein chains break, and the entire system fails catastrophically. The increase of the stress in regimes (III-IV) is reminiscent of a phenomenon referred to as strain hardening. The systems with cracks fail at a slightly lower stress and lower strain than the perfect system. However, all three systems reach a remarkable strain to failure in excess of 135%. This means that the material can be extended by a factor of 2.35 times its initial length without breaking.

Figure 6.7 (a) plots the failure strain as a function of the crack size, for a wide range of values of χ. Interestingly, the failure strain does not vary much among all systems, and even for a crack size of 80%, the material reaches a failure strain that exceeds 130%. This data shows that despite the presence of a flaw inside the protein network, the overall mechanical behavior remains intact and is not severely compromised by the structural imperfection. We find that the maximum stress depends more strongly on the size of the crack as shown in Figure 6.7 (b). However, even the system with 80% crack size still reaches 57% of the strength of a perfect structure without any defects. This performance is unmatched in most synthetic materials, where even small cracks can lead to a reduction of the strength by orders of magnitudes.

To explain this behavior, we carry out a detailed analysis of the deformation mechanism, as shown in Figure 6.8, where the color of the alpha-helical filaments indicates how much it has been deformed (specifically identifying: the elastic regime α – stretching of the alpha-helix without H-bond breaking; the plateau regime β – uncoiling of the alpha-helix through breaking of H-bonds; and the covalent stretching regime γ – the regime where the protein backbone is being stretched). We find that the deformation mechanism of the network is characterized by molecular unfolding of the alpha-helical protein domains, leading to the formation of very large yield regions (Figure 6.8 (b), snapshots II-IV; where the yield regions appear first in yellow and then in red color). These yield regions represent an energy dissipation mechanism to resist catastrophic failure of the system (we thus refer to them as "dissipative yield regions" in the following). Rather than dissipating mechanical energy by breaking of strong molecular bonds, the particular structure of alpha-helical proteins makes it possible that mechanical energy is dissipated via a benign and reversible mechanism, the breaking of H-bonds. Catastrophic failure of the structure does not occur until a very large region of the structure has been stretched so significantly that the strong bonds within and between alpha-helical protein filaments begin to fail. As shown in

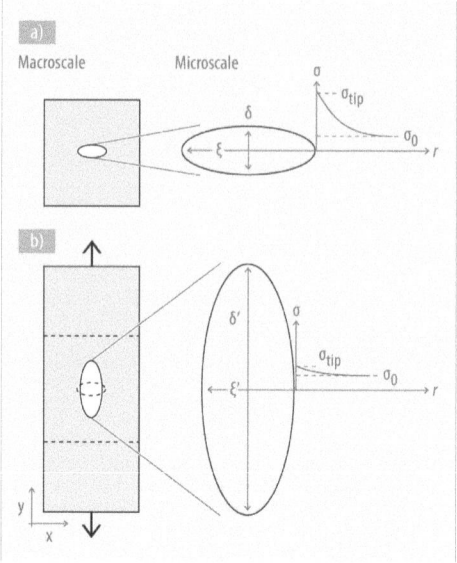

Figure 6.9. Change of the microscopic crack shape as the protein network undergoes macroscopic mode I tensile deformation. Subplot (a) shows shape of the initial crack (an elliptical geometry where the length in the x-direction is much greater than the extension in the y-direction). Subplot (b) shows shape of the final crack before onset of failure, representing an elliptical geometry where the length in the y-direction is much greater than the extension in the x-direction. The plots also indicate the distribution of stresses for both cases (the solution for the stress field is symmetric, but shown here only for the right half). The crack shapes reflect those measured in the simulations shown in Figure 6.8. The initial geometry and crack shape is shown in Subplot (b) (left part) in dashed lines to illustrate the significant transformation.

Figure 6.8 (b) through the highlighted crack shape, we observe that the formation of yield regions enables a significant change of the shape of the crack, from an initial ellipsoidal shape where the longest axis points in the x-direction to an ellipsoidal shape where the longest axis points in the y-direction. This microscopic change of the crack shape induced by the macroscopic applied load is an interesting cross-scale phenomenon with important implications on the failure behavior of the system, as will be discussed shortly.

Figure 6.8 (b) Snapshot IV also indicates that the filaments are relaxed in the x-direction (orthogonal to loading), and are highly stretched in the y-direction (the direction of loading). There is a slight stress concentration at the tip of the crack, as can be seen by the red color indicating stretching of the protein filament's covalent backbone (whereas filaments in the immediate vicinity are strained less). In Figure 6.8 (b) Snapshot V, the entire domain to the right and left of the crack has been unfolded and the backbone is stretched (rupture initiates), whereas the center part of the system has just began to unfold.

In comparison with conventional materials, the protein based material considered here features intriguing fracture properties. To facilitate a systematic analysis we first calculate the fracture surface energy, an important quantity used to quantify the resistance of materials against failure [197]. With W as the energy necessary to permanently break one alpha-helix (through rupture of strong backbone bonds), the fracture surface energy is defined as $\gamma = W/A$, where A is the cross-sectional area associated with a single mesh element. Since $W = 1.63 \times 10\text{-}17$ J (obtained from the integral over the force-displacement curve of an alpha-helical element until breaking of the covalent backbone), and $A = 1.2 \times 10^{-17} \text{m}^2$ (length: 12×10^{-9} m, width: 10×10^{-10} m) the fracture surface energy is given by $\gamma = 1.36 \text{J/m}^2$. This is a value that is comparable to the fracture surface energy of silicon, which features $\gamma = 1.14 \text{J/m}^2$ along the <111> crystal plane, albeit silicon has a much greater elastic modulus of $E = 243$ GPa [204].

According to Griffith's theory (also referred to as the "Linear Elastic Fracture Model", LEFM) - a model often applied to describe fracture of conventional solids - the failure strain for a "central panel" through thickness crack inside a homogeneous material as the one considered here is given by

$$\varepsilon_{0,f} = \sqrt{\frac{2\gamma}{EL\tan\left(\frac{\xi}{L}\pi/2\right)}} = \sqrt{\frac{2\gamma}{EL\tan(\chi\pi/2)}} \sim \sqrt{\frac{\gamma}{E}}, \qquad (6.6)$$

where ξ is the crack length, L is the system width in the x-direction, and $\chi = \xi/L$ (see inlay in Figure 6.6 for the geometry and definition of variables). The scaling of ε with respect to the elastic modulus E and the fracture surface energy γ in eq. (6.6) partly explains the difference in failure strain observed in the alpha-helical protein network compared with materials such as silicon, which typically fail at less than a few percent strain. Due to the much lower modulus (approximately 3 GPa for alpha-helices in regime I-II, versus 243 GPa for silicon) but comparable fracture surface energy, the resulting failure strain is expected to be significantly enhanced in the protein material.

Furthermore, in conventional solids, the occurrence of singular stress concentrations is the reason for rapid catastrophic failure under deformation, as chemical bonds at the corners of cracks are stretched significantly and immensely exceed the deformation and stress imposed at the boundaries of the system. This type of behavior is not observed in the alpha-helical protein network. This is because each of the filaments is able to dissipate a significant amount of energy while they are able to independently stretch without affecting neighboring bonds, as illustrated in Figure 6.8 (b). This is possible since there are no immediate interactions between individual filaments in the network that prevent microscopic rotations and shear (aside from cross-links between filaments present at node points of the mesh). Therefore, these networks do not display a strong stress concentration at corners of cracks. In light of this observation, the relatively low density of filaments with open space between individual constituting elements plays a crucial role in defining their characteristic mechanical properties.

In addition to the particular geometric arrangement in open networks, the properties of individual alpha-helical protein domains are decisive to explain this behavior. The high energy dissipation ability of individual alpha-helical protein filaments is achieved through the particular structure of alpha-helical proteins in combining a large array of small groups of H-bonds, which unfold concurrently in groups of 3-4 at relatively small force levels [200, 205, 206], providing a strongly nonlinear material behavior at the filament level as shown in Figure 6.5 (d).

Notably, the utilization of H-bonds renders the structure self-healing, since H-bonds can reform at moderate temperature (e.g. body temperature) and thereby restore the initial alpha-helical structure even after severe deformation (provided that no strong bonds have been broken). In particular, since in the early relatively flat regime H-bonds are broken that can be reformed rather quickly, the formation of the yield zone that protects the integrity of the structure is effectively reversible upon relaxation of applied load at physiologically relevant time-scales.

We proceed with an analysis of the results in light of fracture models. Figure 6.7 (a) displays an analysis of the failure strain of the system, plotting the failure strain over the relative crack size χ for both, the values measured from the simulation and the predictions from LEFM. The LEFM prediction for the scaling behavior of failure strain versus relative crack size is given by

$$\varepsilon_{0,f} = \sqrt{\frac{2\gamma}{EL\tan\left(\frac{\xi}{L}\pi/2\right)}} \sim \sqrt{\frac{1}{\tan(\chi)}} \sim \sqrt{\frac{1}{\chi}}, \qquad (6.7)$$

suggesting a strong dependence of $\varepsilon_{0,f}$ on χ. However, the simulation results clearly show that the failure strain is largely insensitive to the presence and the relative size of cracks. A power law fit of the form $\varepsilon_{0,f} \sim \chi^a$ to the simulation data reveals that the failure strain $\varepsilon_{0,f} = 1.3275\chi^{-0.0362}$. The prediction based on eq. (6.7) corresponds to a scaling as $\varepsilon_{0,f} \sim \sqrt{1/\chi} = \chi^{-0.5}$. This analysis reveals that the scaling parameter a of $\varepsilon_{0,f}$ versus χ are much different (-0.0362 vs. -0.5), and that the conventional LEFM model fails to describe the failure behavior of this system. A similar analysis is shown in Figure 6.7 (b) for the failure stress, comparing the prediction from LEFM to the measured dependence. Similarly as for the failure strain, the analysis shows that the failure stress remains significantly higher than the corresponding LEFM prediction even at very large relative crack sizes. However, the decay of failure stress is more rapid than the behavior found for the strain.

The behavior of the failure stress on the crack size is investigated further considering earlier solutions for cracks in elastomers [207], which have been developed specifically for the behavior of systems that show strong nonlinear (hyperelastic) and large-deformation elasticity. The maximum strength of the protein network (≈600 MPa) is about 11 times larger than the small-strain elastic modulus (≈56 MPa). This satisfies the criteria for elastic crack tip blunting as discussed in [207]. In agreement with the prediction put forth in [207], large blunting of the tip before failure is observed in the mesoscale experiments (see Figure 6.8 (b)). However, the model for fracture initiation for elastomers put forth in [207] is not directly applicable to our case, since the mechanisms such as void formation or microcracking are not observed in the alpha-helical protein network.

To overcome this limitation we present a simple analysis specific to our case, used here to develop a failure criterion for the alpha-helix protein network. The starting point is the observation that the crack shape significantly changes under the applied load and forms an elliptical geometry before the final stage of deformation associated with the higher stiffness, leading to an elliptical crack shape with a blunted crack tip. A simple approximation of stress fields at a blunted crack tip can be obtained using the Inglis solution for elliptical cracks [208] (see schematic in Figure 6.9 with explanation of variables), where the crack tip stress is given by

$$\sigma_{tip} = \sigma_0 \left(1 + 2\frac{\xi'}{\delta'}\right). \tag{6.8}$$

In eq. (6.8), σ_{tip} (=σ_{yy} at the crack tip) and σ_0 are the stresses at the tip and the far-field respectively, and ξ and δ are the x and y-axes lengths of the elliptical crack shape before failure. Specifically, the parameters ξ' and δ' describe the transformed crack geometry after blunting has occurred through formation of large yield regions mediated by protein filament stretching, but before the final stage of deformation has begun (i.e., before stage II-III shown in Figure 6.6).

Equation (6.8) can be used to make a few interesting points. The equation provides a simple model for the reduction of stress magnification at corners due to structural transformation as discussed above. For an ellipsoidal crack shape where the longest axis points in the x-direction, the ratio $\xi'/\delta' \gg 1$ (Figure 6.8 (b) Snapshot I, Figure 6.9 (a)), the stress at the tip is much larger ($\sigma_{tip} \gg \sigma_0$) than for an ellipsoidal crack shape where the longest axis points in the x-direction, the ratio $\xi'/\delta' < 1$ (Figure 6.8 (b) Snapshot IV, Figure 6.9 (b)), where σ_{tip} is only slightly larger than σ_0. For example, for the geometry shown in Figure 6.9 (a) the initial ratio $\xi/\delta \approx 5$, leading to $\sigma_{tip} = 11\sigma_0$. After the crack shape transformation has occurred, $\xi'/\delta' \approx 0.3$, leading to $\sigma_{tip} = 1.9\sigma_0$, reduced by a factor 5.7.

We may also use eq. (6.8) to develop a simple model to predict the failure stress as a function of the crack size, accounting for crack blunting. By assuming a first order linear relation $\xi' = C_1\xi$ and $\delta' = C_2\delta$ to describe the geometric transformation, we find that

$$\sigma_{tip} = \sigma_0 \left(1 + 2\frac{C_1\xi}{C_2\delta}\right). \tag{6.9}$$

We note that $\chi = \xi/L$ is used to express $\xi = \chi L$, and therefore

$$\sigma_{tip} = \sigma_0 \left(1 + 2\frac{C_1 L}{C_2\delta}\chi\right) = \sigma_0(1 + C\chi). \tag{6.10}$$

The parameters C_1 and C_2 are generally functions of the applied strain. However, noting that failure strain is almost constant independent of crack size (see Figure 6.7 (a)), we assume that C_1 and C_2 take the same value for different crack sizes at failure. It is emphasized that eq. (6.10) contains a constant prefactor $2C_1 L/(C_2\delta) =: C$. The crack will start to propagate when the condition $\sigma_{tip} = \sigma_{max}$ is satisfied, where σ_{max} is the failure strength of a perfect alpha-helical network (since there are no other failure mechanisms such as void or microcrack formation [207, 209] present here). Combining these assumptions with eq. (6.10), we arrive at:

$$\sigma_{0,f} = \frac{\sigma_{max}}{1+C\chi} \sim \frac{1}{1+C\chi}. \tag{6.11}$$

Equation (6.11) is a similar scaling law as proposed in eq. (6.7), but features an unknown parameter C that effectively describes the geometric change of the blunted crack tip under elastic deformation. This parameter can be identified by carrying out a least-squares curve fit for C to the range of geometries considered in our computational experiments, leading to C = 1.102. The results are shown in Figure 6.7 (b), revealing a much better agreement with the simulation data; albeit the model itself is empirical due to the existence of a fitting parameter that must be determined from experimental measurements. However, the model is useful as a constitutive equation to predict the strength of alpha-helical protein networks that can be used in larger-scale simulation methods (e.g. finite element models) to describe the strength behavior of such materials. It might also be used as a design tool to construct systems with optimized values of C that provide less sensitivity to the crack size χ (where possible changes to the geometries at different hierarchical levels could be used as design variables). Possible improvements of the model might be obtained using quantized fracture mechanics models [210] or the development of formulations that account for the specific elastic properties of the system considered here.

It is noted that the definition of "failure" as considered here involves breaking of strong backbone bonds in the network. Under typical physiological conditions this may not occur, since deformation is largely limited to reversible processes at smaller stresses. However, the analysis put forth here provides a worst case scenario to identify the limit of mechanical deformation, which shows that even at modest stresses extremely large deformation can be accommodated without causing any harm to the network integrity. Further, failure modes that may be observed in other systems entail intermolecular sliding of filaments. The analysis discussed here still holds; with the distinction that sliding prevents immediate catastrophic failure of the system but instead leads to the formation of a "plastic zone", formed by the domain in which filaments have begun sliding. This plastic zone provides further resistance against catastrophic breakdown. Indeed, sliding mechanisms have been suggested for intermediate filament protein structures [211, 212].

We finally note that the overall shape of the simulated curve is in good agreement with experimental results (Figure 6.10), published by Fudge *et al.* on hagfish slime threads [22, 58]. In addition to the overall shape, the levels of strain are in good agreement as well, with some deviation at larger strain levels. Simulations and experiments show a change from the flat to the steep regime at about 100% strain and the inset of rupture at about 150% strain (here in experiments sliding sets in). The stress levels diverge due to different geometries of the networks (length of AHs and number of parallel AHs).

6.2.3 Conclusions in light of materials science and biological function

The main result put forth in this Section is that it is due to the particular hierarchical structure and properties of alpha-helical protein constituents that enable the formation of large dissipative yield regions and a severe structural transformation of the crack shape, which effectively protects alpha-helical protein networks against catastrophic failure (Figure 6.8). These yield regions provide a means to dissipate mechanical energy before strong bonds are being stretched and broken, and enables the system to undergo deformation well beyond 130% strain even when cracks are present that stretch of up to 80% of the system size. As a result of formation of dissipative yield regions, the alpha-helical protein networks are largely insensitive to structural flaws, which is reflected in the diminutive influence of the crack size on the failure strain (Figure 6.7 (a)) and the failure stress (Figure 6.7 (b)). This behavior is referred to as flaw tolerance.

The comparison with the scaling behavior predicted from conventional fracture models as summarized in Figure 6.7, and the characteristic failure mechanisms highlighted in Figures 6.8-6.9 illustrates the distinct behavior of alpha-helical protein materials. Thereby each hierarchical level (H-bonds, AHs, network) plays a key role. The dominating unit deformation mechanism of AH protein networks is protein unfolding mediated by continuous rupture of clusters of H-bonds, as shown in Figures 6.8. The detailed fracture mechanism is summarized as follows:

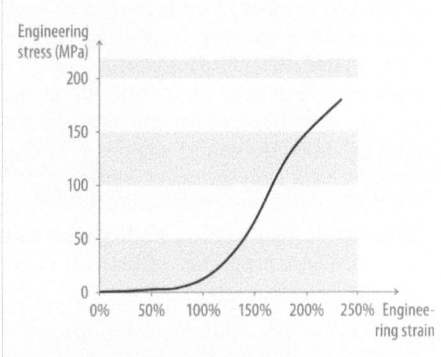

Figure 6.10: Experimental data on the stress-strain behavior of hagfish slime threads [22]. Hagfish slime threads mainly consist of bundles of AH IFs and are thus a good model for comparison with results from mesoscale simulations. Our results show very good agreement with experimental findings regarding the shape of the curve. Both curves have a very flat regime followed by a very steep increase in stress. The change in both curves appears at approximately 100% strain. At about 150% strain flattening sets in, which was suggested as intermolecular sliding. The stress levels mainly depend on the geometry of the bundles/networks.

(a) Initially, the system is loaded in Mode I (tensile load), with the load applied vertically to the long axis of the crack. In solids, this represents the most critical mode of loading with respect to inducing high local stresses in the vicinity of the crack tip.

(b) As load is applied, the protein filaments start to unfold, as H-bonds begin to rupture and the alpha-helical proteins uncoil.

(c) The system elongates in the loading direction, and the shape (morphology) of the crack undergoes a dramatic transformation from mode I, to a circular hole, to finally an elongated crack aligned with the direction of loading (see Figure 6.9). This transformation is caused by the continuous unfolding of the individual proteins around the crack, which can proceed largely independently from their neighbors.

(d) As discussed in the crack blunting model, the elongated crack features very small stresses in the vicinity of the crack. The transformation of the crack shape is thus reminiscent of an intrinsic ability of this material to provide self-protection.

(e) The almost identical strain at fracture (Figure 6.7 (a)) is due to the similar stretching mechanism and unfolding of the proteins at the initial stages of loading. Due to the self-protection mechanism and the related change of the crack shape (that is, the alignment along the stress direction) the crack becomes almost invisible, even if dominating large parts of the cross-sectional area, and has little adverse effect on the overall system performance.

To the best of our knowledge, the studies reported here for alpha-helical protein networks are the first of its kind, providing insight into the fundamental deformation and failure mechanisms of an abundant class of biological materials that feature networks of similar protein filaments. Our results may further explain the ability of cells to undergo very large deformation despite irregularities in the structural makeup of the protein network. This represents an intriguing ability of this class of materials to self-protect themselves against adverse effects of structural irregularities. Avoiding such structural irregularities in the material makeup would require a high energetic cost (e.g. through the need for strong bonding as it appears in crystalline solids). Biological materials solve this challenge by adapting a structure that is intrinsically capable of mitigating structural irregularities or flaws while maintaining high performance, representing a built-in capability to tolerate defects. These properties effectively result in self-protecting and flaw-tolerant materials.

Further investigation could be carried out to provide a more realistic description of the protein network. Our approach does not precisely reflect the specific nanostructure in lamin intermediate filaments as it was designed to provide a rather simple, generic description (see discussion above). Our assumption of a square lattice network of alpha-helical proteins does not accurately reflect the structure of many biological materials, and future investigations could be focused on describing the effects of the differences due to different nanostructural geometries. In these cases, additional levels of hierarchies would enter the structure, resulting in additional mechanisms of deformation and failure beyond those discussed here. For example, sliding between alpha-helical constituents (e.g. in tetramers or larger-scale protein assemblies, see e.g. Figure 6.10) could be an important failure mechanism, which would prevent the immediate drop of the stress to zero as assumed here once this failure mode begins to operate. The possibility of sliding as a deformation mode might explain the slightly lower maximum stress and a deviation from continuous stiffening as seen in experimental analysis of intermediate filament nentworks [22] presented in Figure 6.10 (despite an overall agreement the stress-strain curve shape between experiment and simulation results; where there is a deviation at large stresses). A detailed analysis of the network in dependence of these effects, as well as a quantitative comparison is left to future studies. However, it is pointed out that the mechanisms of self-protection and flaw-tolerance as observed here still hold, because the basic characteristics of the protein network makeup remains similar. The focus on a simple model system as reported here - in the spirit of a model material [83, 201] - provides a clean and well-defined approach to elucidate fundamental mechanisms of failure initiation. If we had focused on attempting to model the particularities of a specific material we would not have been able to identify generic failure mechanisms.

Studies of the mechanical performance of alpha-helical based protein networks as reported here are crucial for advancing our understanding about the deformability, strength and failure behavior of protein materials in general, as well as for our ability to create de novo synthetic nanomaterials for application in biotechnology and synthetic biology. We speculate that our results may also explain the mechanical properties of other biopolymers such as spider silk, where analogous dissipation mechanisms might contribute to these materials' extreme strength and robustness against large deformation. Future studies will be necessary to explore effects specific to these materials. Our findings are also reminiscent of the sacrificial bond concept discussed earlier in the context of bone and other biopolymers [213-216], albeit the sacrificial bond model has not yet been explored in the context of crack-like imperfections and its impact on mechanical performance. Earlier studies of the mineral crystal phase in bone have also pointed out flaw-tolerant behavior, which was linked to nanoscale confinement of mineral platelets [217].

In summary, our analysis, together with studies of single molecule behavior of alpha-helical proteins (see Section 5 for details), improves our understanding of deformation and failure mechanisms of structurally flawed protein networks by providing an integrative model to bring together single molecule properties and larger-scale material behavior through an integrated, consistent multi-scale perspective. A computational approach as put forth here is a promising method that complements experimental investigations. It can also be used to enable a systematic design of materials, by systematically expanding the structural levels on each hierarchy and by designing novel mechanisms beyond those named here. This may one day provide a computational engineering approach similar to what is used in the design of cars, buildings and machines today, applied to the integrated approach that bridges multiple material levels in the design of materials and structures (see Section 7 for further discussion).

7 Summary and discussion

7.1 Summary of main results

The focus of this Thesis was the mechanical behavior of hierarchical AH PMs across several length and time scales. In the first part we discussed our motivation for studying biological materials, and the significance of mechanics in studying them. The studies in this Thesis were undertaken with the protein family of IFs as model system. Their structure, function and mechanical properties were the main focus in Section 2.

Further, we developed a predictive, hierarchical theory that explicitly considers the structural arrangement within the hierarchical protein structure (geometry of HB-clusters as well as their hierarchical arrangement), providing the first rigorous structure-property relationship for PMs. This theory (eq. (4.24) features input parameters solely derived from geometrical arrangement of the HBs (b_i, k_i, l_i), energy parameters E_b and x_b of a HB, as well as the environmental conditions, *i.e.* temperature (captured by the thermal force f_b) and the applied deformation speed v.

This theory allows us to predict the rupture strength as well as the robustness (through studying the sensitivity of strength with respect to microscopic changes, as shown in Section 5.2.6) of a given hierarchical protein structure for different pulling velocities v, and could have similar importance in future nano-scale engineering as now different strength theories on continuum scale (for example those currently used for metals, such as the Tresca model). Importantly, from the theoretical approach, which we choose for deriving the constitutive equations, this theory is valid not only for AH structures, but could also be applied to beta-sheets, beta-helices and other protein structures.

In the second part of the Thesis, we have applied this theory in order to investigate the mechanical effects of point mutations, responsible for laminopathies (case study 1), mechanics of AH proteins over several magnitudes of time scale (case study 2), stutter defects in CCs (case study 3), as well as deformation mechanics on the intermolecular level of two CC dimers (case study 4). The goal was to derive a detailed understanding of the underlying fundamental rupture mechanisms at different length and timescales present in HBMs.

In the first case study, we have explored the effect of a single point mutation on lamin dimer level. The most important result of this study is that a single point mutation does not alter the mechanical properties on the individual dimer level. Our results suggest that the mutation most likely affect larger-scale hierarchical features and properties in the laminar network, such as the dimer or filament assembly or even the gene regulation processes, and thus strongly support earlier hypothesis suggested by experimentalists [34].

In the second case study, within an integrated approach of theory and simulation we have systematically varied the pulling velocity and discovered a change in unfolding behavior of the AH at a pulling velocity of 0.4 m/sec at an unfolding force of 350 pN. Our results prove that the unfolding mechanism at fast pulling rates is rupture of one HB, whereas the unfolding mechanisms at slow pulling rates proceeds by rupture of three parallel HBs (Figures 5.8, 5.9, and Table 5.1).

As of right now, MD simulations are the only means to directly observe these molecular-scale mechanisms, since experiments are still lacking appropriate spatial and temporal resolution. Advances in computing power have enabled us to carry out direct atomistic simulation of unfolding phenomena, including explicit solvent, at time scales approaching a significant fraction of a microsecond. For even smaller pulling rates, reaching the equilibrium state of proteins, we predict a constant force independent of the pulling speed, as entropic effects from conformational changes of the protein backbone are activated and the strength is characterized by a free energy release rate condition. This allows us for the first time to link results from experiments with those observed in MD simulations with a simple, self-consistent model. We have further discovered that 3-4 parallel HBs are the most favorable bond arrangement in light of the mechanical and thermodynamic stability.

In the third case study, we focused on the mechanics of the next higher hierarchical scale, where we have compared the mechanical properties of vimentin CCs with and without the stutter molecular defect [13]. Further, we have performed systematic studies of the pulling velocity. Earlier studies have suggested that the stutter defect plays an important role in filament assembly [53-55]. Our work proves that the stutter also has a significant effect on the mechanical behavior of vimentin dimers under tensile loading. In summary, the stutter: (1) Renders the molecular structures softer, that is, unfolding occurs at lower tensile forces, (2) introduces predefined locations of unfolding, and (3) thus leads to a more homogeneous distribution of plastic strains throughout the molecular geometry. This was illustrated in Figure 5.19, for instance.

In the fourth case study we put forth theoretical predictions of the different deformation mechanisms that appear at the interdimer level. Our analysis provides insight into critical adhesion forces that lead to either CC dimer unfolding or interdimer sliding. Such models are crucial to advance the understanding of biological process like mechanosensation or cell rupture (e.g. cytoskeletal damage) during injuries. Further studies to compare with experimental results could be possible soon, due to recent AFM experiments on IF fibers with high precision [65].

In Section 6 we finally arrive at higher scales, where we put our main focus on the materials science aspects of the studied structures. We first have studied the behavior of hierarchical systems (Section 6.1) and undertook detailed analyses how this architectural feature can lead to materials with high strength as well as robustness, and link them with macroscopically observed properties of stiffness and toughness. We have shown, how with different structural arrangements, different combinations of strength and robustness can be achieved [189]. We have illustrated that the conflict between reaching strength and robustness (at the same time) can be resolved by introducing hierarchies as an additional design variable. This could find broad application in designing future materials.

In Section 6.2 we further have studied the behavior of protein networks as they appear in cells or the nuclear membrane. Hereby we have shown that these networks, consisting of individual AHs on the lowest hierarchical scale are fault tolerant until strains of up to 150%. We have found that the characteristic properties of alpha-helix based protein networks are due to the particular nanomechanical properties of their protein constituents, enabling the formation of large dissipative yield regions around structural flaws, effectively protecting the protein network against catastrophic failure. We have shown that the key for these self protecting properties is a geometric transformation of the crack shape that significantly reduces the stress concentration at corners. Specifically, our analysis demonstrated that the failure strain of alpha-helix based protein networks is insensitive to the presence of structural flaws in the protein network, only marginally affecting their overall strength. Finally, we have shown that the derived stress-strain behavior from our mesoscale simulations agrees well with experimental findings discovered at this length scale.

Each case study reported here was followed by detailed conclusions in light of biological function and a discussion with respect to existing materials engineering concepts. Analogies to other protein materials such as collagenous tissues or BS reach proteins were discussed and differences compared to synthetic materials were analyzed. In the next Section we will broaden the perspective and introduce a theoretical framework, which provides the basis for the analysis of HBMs from a more systematic perspective.

7.2 Nature's hierarchical tool box

PMs, in contrast to more conventional materials such as metals, ceramics or polymers, exhibit a hierarchical design, which allows unifying seemingly contradicting features, resulting in smart, multi-functional and adaptive materials. How is this possible? Providing possible answers and strategies to arrive at scientific insight into this question is a core aspect of this Chapter.

Even though PMs lead to vastly complex structures such as cells, organs or organisms, an analysis of their composition reveals simple underlying mechanisms that can be classified into two major categories. Some of the structural features materials are commonly found in different tissues, that is, they are highly conserved. Examples of such universal building blocks include alpha-helices, beta-sheets or tropocollagen molecules. In contrast, other features are highly specific to tissue types, such as particular filament assemblies, beta-sheet nanocrystals in spider silk or tendon fascicles [15]. These examples illustrate that the coexistence of universality and diversity through hierarchical design – in the following referred to as the universality-diversity paradigm (UDP) – is an overarching feature in protein structures.

This paradigm is a paradox: How can a structure be universal *and* diverse at the same time? In PMs, the coexistence of universality and diversity is enabled by utilizing *hierarchies*, which serve as an additional dimension, enlarging the 3D or 4D physical space.

The particular focus of this Chapter is the discussion of an improved understanding of the relations between hierarchical structures, mechanical properties and biological function of PMs [218]. Up to now several different aspects related to that were presented in previous chapters. Now we ask the questions: How does the hierarchical arrangement influence the mechanical properties such as strength and robustness? Which path follows Nature in adapting materials that unify multiple functions in one 'material', finally enabling life under the vast variety of environmental requirements and its continuous changes? Answering these questions will allow engineering future bio-inspired and bio-mimetic, possibly in-organic structures, based on millions of years of 'experience' collected by Nature.

Virtually all PMs create multi-functional and highly robust structures that - without wasting resources - arrive at satisfactory solutions [15, 185, 219, 220]. Moreover, most biological materials feature a decentralized organization [180, 221, 222], wherein self-organization, self-regulation, and self-adaptation govern the formation, reformation and repair or healing at multiple time and length scales.

Synthesizing these structures in novel materials in a controlled fashion represents a great opportunity that can be tackled in the next decades.

7.2.1 System theoretical perspective on biological structures

Hierarchies in biology and biological materials

Hierarchical systems have already been observed previously in many different (non-) biological areas. In system theory, a hierarchical system is defined as a composition of stable, observable sub-elements that are unified by a super ordinate relation [123]. Thereby, lower level details in a complex hierarchical system may influence higher hierarchical levels and consequently affect the behavior of the entire system. Therefore, the interactions between different hierarchical levels or, equivalently, hierarchical scales are the focal point in system theory based concepts of hierarchical systems.

Importantly, averaging over one scale to derive information for the next higher scale is generally not feasible. This is because either an insufficient number of sub-elements is present [122], or because a particular piece of information may be forfeited that might be crucial for the behavior several scales up [123].

One of the best understood hierarchical systems is the 'hierarchy of life', where cells, organs, organisms, species, communities, and other entities are put together in an inclusive hierarchical relation [123]. However, in the hierarchy of life a cell is the smallest hierarchical subunit. In the last decades, several additional subunits ranging from cellular to the atomistic level have been discovered, including protein-networks and individual proteins, reaching down to the scale of AAs.

The discoveries on small scale gave among other rise to a new discipline: The science of system biology, where the focus lies on understanding a system's structure and dynamics, such as traffic patterns, its emergence and control, or signaling cascades [180]. A terminology was adapted from system theory into the system biological context, whereof the most relevant terms for HBMs are listed below and summarized in Table 7.1.

Robustness and complexity

Biological materials and systems are critical elements of life. That is why it would be very harmful if the failure of a single component would lead to a catastrophic failure of the whole system. Thus a major design for biological materials is robustness towards the failure of single components or changing environmental conditions, in other words, the maintenance of some desired systems characteristics despite any fluctuations.

Kitano classifies robustness of biological systems in three ways [179, 180]: (i) Adaptation, which denotes the ability to cope with environmental changes, representing

Hierarchical systems	A hierarchical system is a system composed of stable, observable sub-elements that are unified by a super ordinate relation. Thereby, lower level details affect higher levels and thus the overall system behavior.
Complexity	Complexity arises in systems that consists of many interacting components and leads to emerging nonlinear behavior of a system.
Robustness	The following three classes of robustness are suggested to be relevant for biological system: (i) adaptation to environmental changes (external perspective), (ii) parameter insensitivity (internal perspective) and (iii) graceful degradation after system failure rather than catastrophic failure.
Protocols	Protocols are rules, which are designed to manage relationships and processes, building the architecture and etiquettes of systems. They are linking different elements as well as different hierarchies in a system.
Optimality and perfect adaptation	It is commonly believed that random changes in (biological) systems, supported by protocols give rise to new structures and features, leading to a continuously improved performance of a system, which finally results in perfect adaptation of the system and optimal fulfillment of a required function.

Table 7.1: Summary of a selection of system theoretical terms and concepts used in this Thesis, used here to describe the overall behavior of biological systems.

the 'external perspective', (ii) parameter insensitivity, representing the 'internal perspective' of robustness, and (iii) graceful degradation, reflecting the characteristic slow degradation of a system's function after damage, rather than catastrophic failure.

However, robustness has its costs. One mean in realizing robustness are redundancies, *i.e.* many autonomous units carry out identical function. Examples are multiple genes, which encode similar proteins, or multiple networks with complementary functions in cells [179]. Higher degrees of complexity are partly believed to represent another kind of costs of robustness, following the assumption that biological systems are results of a trade off between robustness and internal simplicity [178, 220, 223-226].

Nevertheless, there is yet no consensus, if biological systems are complex or not. As shown above, parts of the biological community believe that complexity is necessary for robustness and thus essential for biological systems [223]. Others believe that 'coherent' or 'symbiotic' are attributes that describe biological systems in a better way than 'complex' [179]. A third group of scientists finds that biological systems are much simpler than we assume, given the fact that cells evolved to survive, and not for scientists to understand [220, 221].

Simplicity, modularity and protocols

How does Nature treat the conflict between robustness and simplicity, resulting in a controlled degree of complexity? Applications of a limited number of universal building blocks, network motif or modules seem to be the path to success [220, 221]. Alon illustrates this simplicity on gene-regulation networks, which are build out of only a handful network modules [221]. But modularity does not only occur on the gene level. It plays an equally important role from base pairs and AAs to proteins, from organelles and membranes to pathways and networks, and finally to organs and organ axes. Additionally, even complex processes, such as protein folding, were shown to be much less complex than expected for a long time [220, 222].

An additional source of simplification in biology is the strong separation of time scales for different processes, e.g. the production of proteins takes place on the time scale of minutes where the chemical modification of protein networks is realized only within seconds [221].

Finally, Wolfram has indicated in his studies with simple programs that the degree of complexity in biological systems can be achieved through simple rules and elements [227]. Another word for rules is *protocols*, which are designed to manage relationships and processes, building the architecture, interfaces and etiquettes of systems. Thus, abstractions such as gene regulation, covalent modifications, membrane potentials, metabolic and signal transduction pathways, action potentials, and even transcription-translations, the cell cycle, and DNA replication could all be reasonably described as protocols [220]. However, the simplest protocols, being the ones realized on the atomic scale, are force-fields describing the covalent and non-covalent interactions, such as HBs, Coulomb interactions or van der Waals interactions.

In general, specific protocols describe the interaction between elements as well as between different scales in a hierarchical system. A good protocol is one that supplies both robustness and evolvability. Therefore, successful protocols become highly conserved because they both facilitate continuous evolution but are themselves difficult to change [220, 225].

Perfect adaptation and optimality, evolvability and recreation

The standard Neo-Darwinian theory of evolution is based on the idea that random genetic changes, coupled with natural selection, will result in progressive transformation of form, which can give rise to new structures and functions in organisms [228]. Protocols support this process of adaptation by activating 'algorithms', which optimize fitness functions. The result of this optimization process is perfect adaptation towards different structural and environmental requirements [220, 223-226].

Perfect adaptation means maximal efficiency, which leads one to conclude that each element has its own place in a biological system. Once this element is taken out while its function is still activated, a new element is created, which newly will fulfill this particular function. The same happens, if new functions appear. The evolving niche is closed by new elements or existing but adapted ones. This mechanism of adaptation and (re-)creation was proved for macroscopic biological systems (e.g. fruit fly species on Hawaii) [229], or for an efficient organization of collagen on another length and time scale. This example refers to the fact that collagen is only present if the mechanical load is applied and otherwise degrades, which leads to dynamical adaptation of collagen networks suitable for a particular load condition. Consequently, why should perfect adaptation and efficiency not be generally governing micro- and nanoscopic structures and processes?

As demonstrated, system theory and system biology provide first significant insight in the properties of biological system. However, there is up to now, to the best of our knowledge, no theoretical paradigm that describes multiple perspectives of HBMs in an integrated manner. This has prevented researchers from fully understanding the structure-property relationship of HBMs, and has limited applicability of concepts found in HBM in technological applications.

7.2.2 Generic paradigm: Universality and diversity in hierarchical structures

As shown in the previous Section, and as it will be illustrated later in the example of IFs, HBMs can be a great source of scientific and technological inspiration. It appears as if the hierarchical design is an essential feature in Nature in general, enabling to unify synergistically contradictive dimensions (e.g. universal/diverse, global/local), resulting in multi-functional biological materials with highly adapted (e.g. on the assembly level), yet robust (e.g. individual alpha-helices) properties.

However, up until now a theoretical framework that enables to address relevant questions in HBMs systematically within a unified multi-perspective approach has remained elusive. With the generic UDP presented here, as summarized in Figure 7.1, we hope to close this gap.

Unifying strength with robustness through hierarchies

Csete and Doyle claimed already that optimality and robustness are most important to biological systems [220]. But how does this refer to HBMs? We believe that from the mechanical point of view the parameter 'strength' has to be optimized and thus replace optimality in this context.

The properties 'strength' and 'robustness' are contradicting properties that can not be combined within a single scale of 'traditional' materials. This was shown already in chapter 6.1.2. Many materials and structures engineered by humans bear such a conflict between strength and robustness; strong materials are often fragile, while robust materials are soft. Fragility appears due to the high sensitivity to material instabilities such as formation of fractures [83, 201]. Consequently, if extreme conditions are expected, only high safety factors, requiring more resources can guarantee the strength of engineered materials [230].

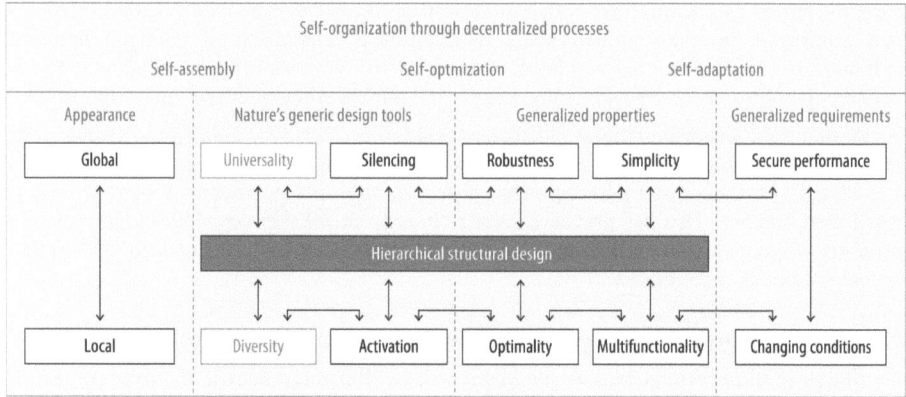

Figure 7.1 Universality-diversity paradigm (UDP) for HBMs. HBMs consist of the hierarchical design on the one hand, and the decentralized organization of processes on the other hand. Thereby, the processes are characterized through decentralized self-organization, including but not limited to: self-assembly, self optimization and self-adaptation, which can be realized through hierarchies, an additional dimension, extending the 3D/4D physical space. This guarantees a synergized unification of seemingly unlinkable attributes of Nature's tool box, which is necessary to realize 'generalized properties', which are required to fulfill specific functions as set up by the environment, among other by the need to survive.

This example illustrates that it is impossible to combine strength and robustness at a single scale; instead, structures with multiple scales must be introduced, where universal and diverse patterns are unified hierarchically. In these structures, universality generates robustness, while diversity enables optimality. Materials like bone, being a nano-composite of strong but brittle and soft but ductile materials, illustrate this unification of components with disparate properties within a hierarchical structure [231]. A detailed analysis in Section 6.1 has shown how universal building blocks can lead to different mechanical properties purely through structural rearrangements possibly maximizing both strength and robustness.

Obviously, extreme mechanical conditions (such as high loading rates and deformations) have to be sustained in Nature under limited access to 'building materials', which make the combination of strength and robustness imperative for existence. Therefore, materials found in biology are very efficient due to robustness, and thus capable of minimizing waste of resources that otherwise appears from high safety factors. A simple calculation illustrating that was undertaken in Chapter 6.2.3. We address this phenomenon explicitly in Chapter 7.3.2.

Certainly, optimality might also appear in a non-mechanical sense, such as optimized thermal, electrical or energy organization and conductivity. However here we focus solely on mechanical and directly related properties.

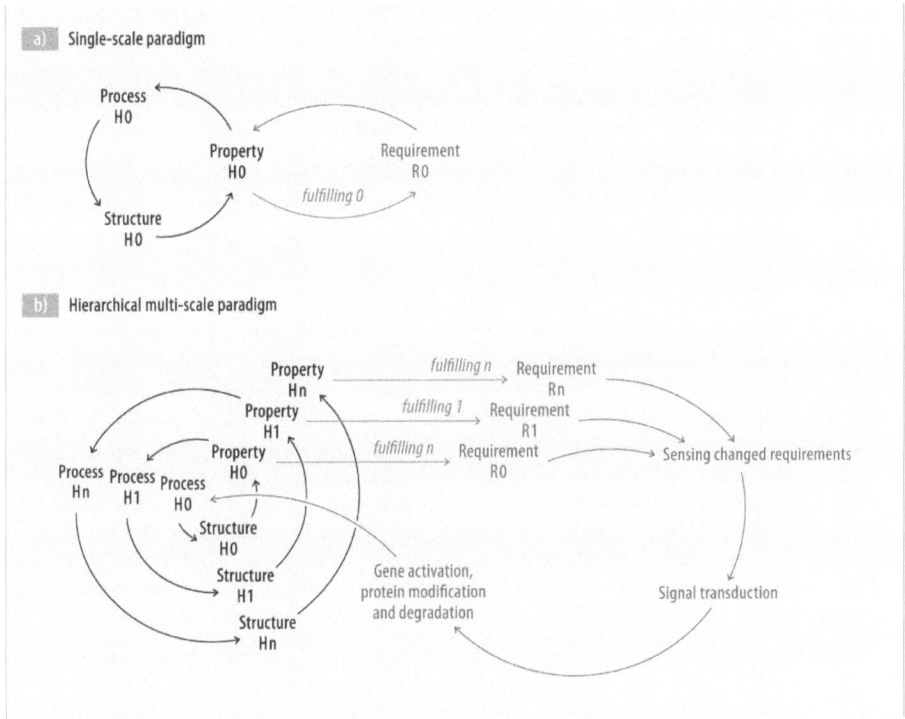

Figure 7.2: In biological materials hierarchical structures, decentralized processes, material properties and environmental requirements, are brought together in mutual completion. Subplot (a) illustrates the traditional paradigm in materials science where process, structure and property build the "magic" triangle on a single hierarchical level. Subplot (b) illustrates the paradigm for hierarchical (biological) materials. In contrast to the traditional paradigm, relations between "external" functions/requirements and "internal" properties exist on several scales resulting in multifunctionality. Further, as requirements are consistently changing over time (e.g. changing loads, changing environment), continuous adaptation is necessary. Though, in addition to multifunctionality, robust feedback loops, resulting in smart signaling chains allow decentralized self-organization. Consequently, in HBMs level-specific properties (H i) do not only fulfill the required functions, but also initiate the decentralized processes on the next hierarchical level (H i+1), and thus generate the structures on this level (H i+1).

Controlling properties through silencing and activation

Particular features of HBMs are silencing and activation mechanisms that act on different scales. These mechanisms represent a set of 'tools' that provide the ability for local optimization while simultaneously guaranteeing global robustness. Examples of this phenomenon were presented in Chapters 5.1 and 5.3.

Robustness is guaranteed when differences or changes that appear at the lower hierarchical scale do not influence higher scales (e.g. alpha helices), that is, expressing *silencing* (robustness in the sense of parameter insensitivity), which allows a global application of this particularly stable feature.

In contrast to that, if an element has great potential to *activate* larger scale properties, that is, its changes appear 'nonlocal in scale', its application is not 'safe' and conservation is unlikely. Given that systems, which are robust against common or known perturbations can often be fragile to new perturbations [179, 223, 224], it may not be surprising that these 'unsafe' features are extensively applied whenever (self-) optimization and continuous adaptation are necessary (robustness in the sense of environmental adaptation). This aspect might explain why universal patterns are more often found on a lower hierarchical level, whereas diversified patterns appear at higher scales.

Remarkably, the question of local versus global changes seem to be relevant not only for HBMs, but also for other processes, such as gene regulation [232], illustrating that this may be an overarching paradigm in biology.

Unifying multifunctionality with controlled complexity

Engineered structures and systems in our days (e.g. airplane, car or building) now reach a similar degree of multifunctionality as biological systems [220]. However, many engineered multi-functional structures have an uncontrollable degree of complexity, as a multitude of distinct elements are combined on a single or few hierarchical levels. Human organizations, in contrast, realize multifunctionality through hierarchical, but yet highly complex structures. Approaches to create self-organized systems, such as the internet, which are based on a standardized 'protocol' are simple, yet fragile. This fragility is observed when bugs in the software appear, or viruses and spam (or certain types of overloads) spread very rapidly, without noticeable resistance. This is because these viruses utilize mechanisms that are compatible with the particular protocols in the network, and decrease the efficiency of or even knock out entire networks [233, 234].

In contrast to that, Nature follows a different path. Here, multifunctionality is created through hierarchically combining universal and robust patterns on selected levels/scales with diversified, but highly adapted elements on others. This results in robust and multi-functional, yet simple systems. Thereby, the level of complexity is kept under control, making the structure as whole more efficient. Instead of reinventing new building blocks, universal patterns and protocols (e.g. specific kinds of chemical bonding) are utilized and 'internal degrees of freedom' that arise from lower scales are kept. These degrees of freedom are 'forwarded' to higher scales, where their application is necessary, for instance for biological function. This concept of *silencing* enables to adapt systems without significantly changing them, and appears to be a universal trait of biological systems.

Decentralized processes: Breaking the symmetry

Remarkably, in contrast to Nature's structural design, which is dominated through hierarchies, Nature's *process* design is dominated through decentralization and self-organization, represented through self-assembly, self-regulation, self-adaptation, self-healing and other processes (see Figure 7.1).

Interestingly, the decentralized processes seem to lead to a multi-scale perspective in time, where different time scales are covered, ranging from nano-seconds for creation of individual HBs, over minutes for assemblies and rearrangement, to eons for adaptation and optimization. The separation of processes through different time scales makes also sense from the biological point of view, as this increases both simplicity and robustness (see above).

Linking structure and process

As illustrated in Figure 7.2, we believe that in biological materials, hierarchical structures, decentralized processes, material properties and environmental requirements, are brought together in a mutual completion.

In contrast to the traditional paradigm in materials science, relations between 'external' functions/requirements and 'internal' properties exist on several scales resulting in multifunctionality. However, since the requirements in biological systems constantly change (e.g. changing loads, changing environment, diseases) on several time and length scales, in addition to multifunctionality, robust feedback loops are required to enable decentralized self-organization and self-optimization.

This clearly shows that in HBM structure and processes are merged intimately.

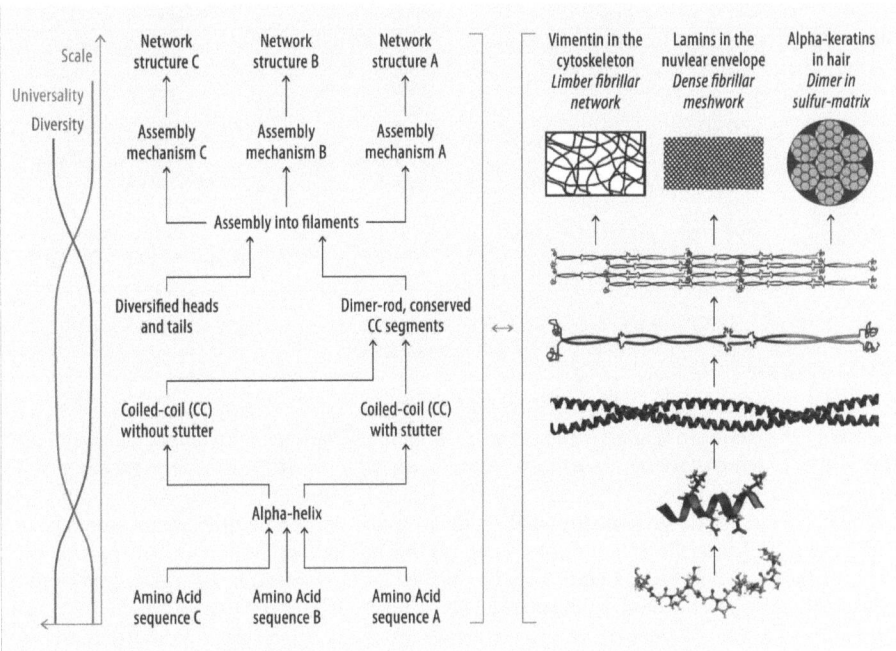

Figure 7.3: Hierarchical biological materials, here exemplified on IFs, are governed through interplay of universal and diverse patterns, which, combined with silencing and activation are unified over multiple scales. This enables to forward the information, completely coded at the lowest scale (AA sequence), safely by means of silencing through intermediate scales (alpha helix, CC) up to larger length-/hierarchical scales, where they are activated in order to fulfill specific requirements. The scale-characteristic patterns are illustrated on the right side.

7.2.3 Application of UDP on IFs

In the previous Section we have introduced the UDP. In this Section we will exemplify on IF proteins, how universality and diversity, silencing and activation are combined in a hierarchical structure, building materials with multiple, scale specific functions, which on their part are combined with scale specific processes. These findings are summarized in Figure 7.3.

Silencing and activation in IFs

The lowest level of hierarchy encodes the structure of these proteins in the sequence of AAs. This is reflected by the fact that each IF type has a distinct AA sequence. Intriguingly, the differences at the lowest hierarchy do not obligatory influence the immediately following hierarchical level. This can be verified since all IFs feature the AH motif, despite differences at the AA sequence level and/or differences at larger scales.

However, moderate effects can be observed at the dimer level. Herein, for example AA

IF type	Found in ...	Biological and physiological functions		
		Protein level	Filament level	Cellular/ network level
Vimentin	Cell's cytoskeleton	Cell signaling mechanisms, associated protein organization	Responsible for location, shape and stability of cell organelles, protein targeting processes	Security belt' of the cell
Keratin	Cytoskeleton, hair, nails, hoofs	Protein synthesis, cell signaling mechanisms, associated protein organization	Cell pigmentation, organization of cell organelles	Cell growth, wound healing, locomotion, prey procurement
Lamin	Nuclear envelope	Signaling mechanisms, mechano transduction, chromatin positioning	Gene regulation and transcription, chromatin positioning	Protection of the chromatin, involved in cell mitosis

Table 7.2: IFs are remarkable due to their diverse appearance in organisms, where they satisfy multiple functions at different hierarchical levels. Interestingly, the elementary building block of all kind of IFs is identical - the universal AH CC protein motif [9-11].

inserts in the periodic heptad repeat lead to a local uncoiling of the super helix (creating the stutter), which effects assembly process as well as the unfolding mechanics (as discussed in Section 5.3) [13, 34]. Even if all IFs commonly show an assembly into filaments, lower scale differences (that is, for instance the AA sequence and stutter) affect the pattern and process of assembly, such as the number of proteins per filament cross area, or the way dimers associate (head-to-tail in lamins versus head-to-head and tail-to-tail in vimentin) and others. In particular the differences on the filament level are of vital importance, as they influence the properties on the network and the super-structural level, which are dominated but not limited to mechanical functions, such as strength or energy dissipation. In the following examples links between the hierarchical design and the resulting multiple functions and processes are shortly brought.

The multiple functions of the different IF types are also summarized in Table 7.2.

Multifunctionality of IF proteins

Vimentin networks in the cytoskeleton act mainly as the 'security belts' of the cell [32, 235]. Due to their architecture, the flexible networks are very soft at small deformations and pulling rates, leading to 'invisibility' and non-resistance during cell movement.

Contrarily, a very stiff behavior is observed at high deformations and high deformation rates, ensuring their function on the cellular as well as on the tissue level [33].

However, recently additional functions have been found on the sub-network level (filament level), which are still but less mechanical. Vimentin networks were proved to be not only responsible for the location, shape and stability of cell organelles (e.g. mitochondria or golgi), but also for their function as well as for the protein targeting process [9]. And yet other function exist on the molecular level, consisting of different regulation mechanisms such as cell signaling (e.g. transcriptional effects, mechano transduction), or associated protein organization (e.g. plectin, chaperones) [9, 218].

Keratin networks in skin tissue, hair, nails and hoofs, Representing one of the main cytoskeletal components in skin epithelia cells [236, 237], fulfill similar structural functions as vimentin, which are, protecting cells from mechanical and non-mechanical stresses, enabling cell signaling, or organizing cell organelles and keratin associated proteins. But that is by far not all.

Additionally, evidence was brought that keratins are responsible for several skin cell specific processes such as cell pigmentation (hyper- or hypo-pigmentation of the skin due to keratin mutations), cell growth, protein synthesis and wound healing (controlled through keratin signaling chains) [237, 238], bringing prove of 'perfect adaptation' of this protein structure towards additional functional requirements on the surface of organisms.

Even more fascinating is that α-keratins also build the main component of hair, nails, hoofs and claws (and β-keratins are the main component of the even harder materials such as turtle shells or bird beaks), where micro- and macro fibrils are embedded in a sulfur rich matrix [239-242]. This enables to provide macroscopic mechanical resistance for locomotion or prey procurement.

As illustrated above, the case of *lamin networks* is slightly different than the two previous examples, because lamins are associated with the inner nuclear membrane of cells, where they provide a dense and resistant network against compression [243, 244]. This architecture enables them to realize their mechanical function, which is to protect chromatin in the nucleus from mechanical load. Diseases related to mutations in lamins, such as skeletal or cardiac myopathies (e.g. Emery-Dreifuss muscular dystrophy), lead among other to uncontrolled rupture of the nuclear envelope [75].

However, similar to the previous cases the role of lamins is not purely structural. In addition to the structural hypothesis, the 'gene regulation hypothesis' is gaining broader acceptance, which gives lamins a key role in the organization of DNA as well as in the gene transcription process (see also Section 2.5) [69, 75, 153, 245]. Further, lamins are suggested as one key element in the signaling chain, forwarding signals from the cell-membrane to the DNA, where a specific response is triggered [69].

Coexistence of universality and diversity

The case of IFs illustrates how hierarchies are applied in order to unify universal robust elements (AHs) and highly diversified and optimized patterns (specific head-tail domains, network architecture etc.). As shown in this example, nanoscopic modifications (e.g. AA sequence) do not always influence the properties at the next hierarchical layer, but those of one or more hierarchical layers above. It appears as if specific functional requirements at several higher scales are 'forwarded' to lower scales, where modifications are implemented. Through this mechanism biological materials are not only multi-functional but are further continuously adapted to the required scale-specific processes, with the goal to fit the diverse required functions in the best possible way, thus ensuring optimality.

7.2.4 Significance of UDP for understanding of HBMs

Ten years ago a systems engineering framework has given birth to a revolutionary approach in the form of quantitative conceptual design of materials [246]. In this framework a resonant bonding between the science and the engineering of materials was created, in which the deductive cause-and-effect logic of science has been combined with engineering approaches, while inductive goal-and-means relations of engineering influence science approaches. This advancement, combined with a hierarchy of design models and theories over several length and times scales, allowed a bottom-up design of desired materials [246].

Now it may be the time to create this type of engineering approach for biological materials, allowing to utilize the insights from science in engineering solutions and *vice versa*. Thereby a structured classification of relevant material properties and mechanisms found in biological materials are crucial in enabling their systematic understanding. A first step in this direction was achieved in the previous sections through the development of the UDP. Appropriate models for hierarchical subsystems, as well as the corresponding links between them need to be developed and validated in the next steps.

Nature apparently followed the path of trial and error. Thereby a radical discrimination between the essential and the nonessential properties regarding the necessary function appeared in the evolutionary process over millions of years, where most of the nonessential 'waste' was eliminated. This resulted in very strong cause and effect mechanisms and finally resulted in a handful of building blocks (20 AAs, secondary structures, etc.) arranged in a hierarchical manner. In biology, different material property requirements are fulfilled just by adopting the hierarchal architecture of these building blocks rather than by inventing new building blocks. Examples, how different structural arrangements can unify disparate mechanical properties were brought in Section 6.1.

A system-theoretical framework as the one described above will allow us to study and understand these cause-effect-mechanisms. Eventually, one day we may have a construction kit that consists just out of a few nanoscopic elements such as carbon nano-tubes (CNTs), nanowires, and protein domains. Paired with the understanding of cause-and-effect mechanisms over several hierarchies, we might be able to "compute" material designs for specific requirements directly out of these simple building blocks. A reasonable comparison for this building kit exists in civil engineering. Here out of bricks, construction steel, concrete and glass a vast variety of buildings for work, habitation, leisure, light harvesting and others can be achieved directly through different architectures.

7.3 Learning from Biology, bio-inspired hierarchical structures

In this Thesis we have presented many mechanical aspects of biological materials. Different mechanisms (and their interplay) such as the rupture of one vs. three HBs, unfolding vs. backbone stretching, stochastically vs. thermodynamicly governed unfolding (all three for AHs and CCs), sliding vs. protein unfolding (tetramers), silencing vs. activation or fault tolerance were observed and analyzed theoretically, by applying basic physical concepts. Here we discuss these findings in a broad materials science concept.

7.3.1 Interpretation of HBMs in light of materials science concepts

Here, different observed mechanisms were described theoretically. The theoretical progress in understanding PMs at the atomistic scale will enable us to understand, and eventually to exploit the extended physical space that is realized by utilization of hierarchical features. By utilizing a bottom-up structural design approach, the hierarchical, extended design space could serve as means to realize new physical realities that are not accessible at a single scale, such as material synthesis at moderate temperatures, or fault tolerant hierarchical assembly pathways [229]. These traits may be vital to enable biological systems to overcome the intrinsic limitations, for example those due to particular chemical bonds (soft) and chemical elements (organic).

The increased understanding of the hierarchical design laws might further enable the development and application of new organic and organic-inorganic multi-featured composites (such as assemblies of CNTs and proteins or polymer-protein composites [190-192]), which will mainly consist of elements, appearing in our environment in an almost unlimited amount (C, H, N, O, S).

These materials might consequently help to address human's energy and resource problems (e.g. fossil resources, iron, and others), and may enable us to manufacture nano-materials, which will be produced in the future by techniques like recombinant DNA [1, 2, 247] or peptide self-assembly [248-250], techniques where the boarders between materials, structures and machines vanish.

7.3.2 Robustness allows reducing safety factors

Robustness plays an elementary role in biological materials. As we have shown in our case studies, a delicate balance between material strengthening and weakening exists in biological materials, providing robustness through geometrical changes, patterned at nano-scale (stutter). Our results support the hypothesis that Nature seeks to provide robust mechanical function in biological materials. For example, the assembly into CCs does not only strengthen the material, but it also allows to create structural features that direct towards a more controlled unfolding and uncoiling behavior compared to single AHs (the stutter is a feature that can not exist based on a single AH protein; it is a property that emerges at the level of a dimer). Slight reduction in mechanical strength is sacrificed in order to obtain a robust and controlled unfolding behavior, independent of the loading rate, following the rule 'safety first'.

Understanding Nature's realization of robustness might inspire future design of synthetic materials, since up until now, due to missing robustness, materials and structures engineered by humans typically demand very high safety factors, which guarantee their function even under extreme conditions. For instance, a structure like a bridge must be able to withstand loads that are ten times higher than the usual load. This is necessary since these structures are very fragile due to their extremely high sensitivity to material instabilities such as cracks. In contrast, as shown for AHs, materials found in biology are often very efficient due to robustness, and thus capable of minimizing waste of resources that otherwise appears from high safety factors. Biological materials must be able to sustain extreme mechanical conditions (such as high loading rates and deformations) under limited access to 'building materials', which make the combination of strength and robustness imperative for existence. We have shown the appearance of such a phenomenon in Section 6.2.

7.3.3 New opportunities for scientists and engineers

The detailed analysis of HBM could contribute to different scientific disciplines, such as science of fracture, materials theory, genetic research (e.g. the hierarchical three dimensional folding of the DNA, indicating the link between structural organization and function [251]), and might further contribute to the understanding of which driving forces in Nature create HBMs.

Additionally, the hierarchy-oriented approach (integrated in the UDP) might integrate different scientific strategies (e.g. macroscopic [25-27] versus nanoscopic [217, 252-254] approaches in understanding fracture of bone), through the holistic consideration of problems, using the concept of coexistence of universality and diversity at different scales and application of both through fundamental design laws.

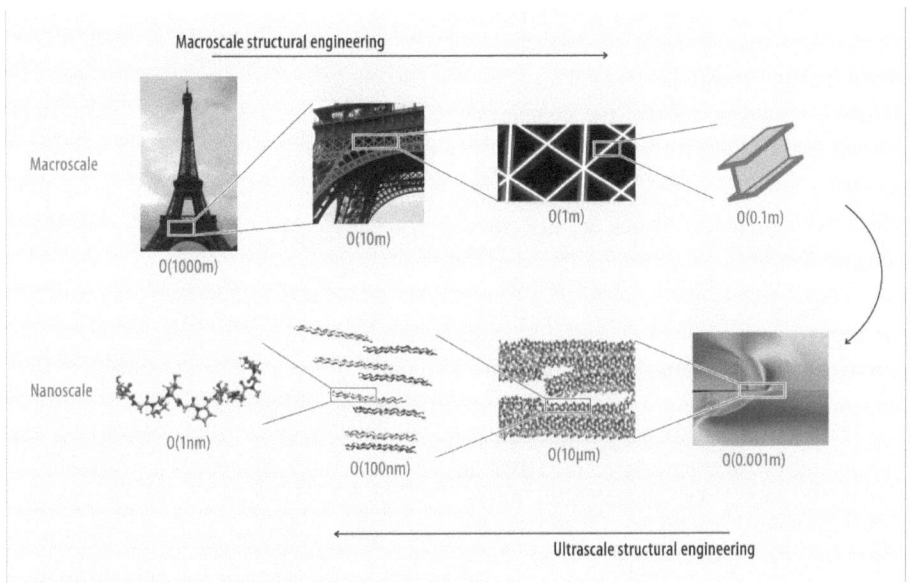

Figure 7.4: Merger of structure and material in engineering design. The long term impact of this work is that it could extend our ability to perform structural engineering at macroscale, to the ultimate scale, the nanoscale. Opening the material scale as design space for new material development may open endless possibilities for development of robust, adaptive, active and 'smart' materials.

Elucidation of the controlling factors in achieving universality and diversity, as well as the understanding of its impact on robustness and optimality, will lead to a paradigm shift that emphasizes on simultaneous control of structural features at all length scales and hierarchies (Figure 7.4). Engineers might be able to design smart sensor-actuator networks on nano-scale, which will enable chemo-mechanical transduction, leading to self-organization and adaptation to the environment. These networks will be part of micro-machines, which will be able to perform complicated tasks in a robust and secure way. These machines, being part of higher order structures, will enable through their high level of cooperation self-adaptation, self-strengthening and self-repair.

Further, a detailed understanding of HBM and the generation of appropriate HBM-models from cells and extra cellular tissues with a particular focus on the link between structure, function and process, as well as scale interaction and connection, will lead to an immense progress in the rising field of nano-medicine and affect other industries such as pharmaceutical and cosmetic industry. These models will for example help to improve drug delivery systems or *in vivo* tissue repair processes.

In more general terms, researching hierarchical PMs (through the eye of the UDP) will provide a fundamental understanding of the question of repeated use of templates versus the making of new structures or components and its assembly in hierarchical structures. This might inspire future product design as well as manufacturing and assembly strategies, not limited to the nano-scale. For example, using universal patterns to the fullest extent and creating diversity at the highest hierarchical level, in order to match client-specific requirements, could reduce production costs, delivery times and increase product quality.

It has been suggested that the complexity of engineered systems is converging with the one of biological systems. For example, a Boeing 777 has 150,000 subsystems and over 1000 computers, which are organized in networks of networks [220]. Consequently, a better understanding of how nature designs and manages complexity will enable to hold engineered complexity under control or even reduce it.

7.3.4 Impact of HBMs on other disciplines

An extended understanding of the HBM at nanoscale paired with hierarchical multi-scale modeling (see Figure 7.5 and Figure 7.6) and petaflop computing may have additional beneficial implications beyond scientific and engineering disciplines, such as creation and optimization of infrastructure networks (e.g. energetic, communication), organization or transportation systems, and many others.

Similar to engineered systems, new ideas and approaches will reduce the complexity of these structures by simultaneously increasing robustness and adaptability as well as flexibility – both crucial attributes in today's quickly changing world. Thus,

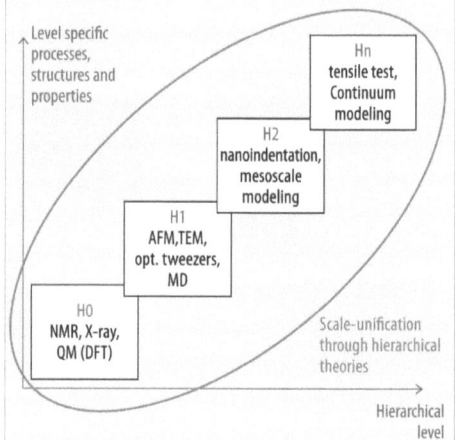

Figure 7.5: Each hierarchical level in biological materials has its specific processes, structures and properties. In order to gain a detailed understanding of HBM on each scale as well as of the interaction between different hierarchies, theory, simulation and experiment will have to work together extensively. While simulation and experimental techniques are mostly limited to a certain length scale and therefore to a few hierarchical levels, new theories, based on information and knowledge from different hierarchical levels will describe the fundamental cross scale relations and thus provide explanations for observations on different scales. This illustrates exemplarily the importance of collaborations between different approaches (theory, experiment and simulation) and disciplines (biology, medicine, physics, chemistry, materials science, computer science, and others) for future progress research.

adaptive organizations and networks will lead to a better performance and consequently to a continuous economic growth, while the robust way these systems operate will increase the well-being of employees and citizens.

Significant impact could also be achieved in urban area design [255]. Hierarchically organized regions and cities, where the functional links between the sub-elements are inspired by biology, could for example solve the traffic problem in big cities or dramatically slow down the spreading speed of epidemics in populous areas [256].

Finally, at a larger time scale, national and international political systems, inspired through robust hierarchical biological structures, could become more efficient in their tasks, e.g. through protecting human rights over several levels simultaneously (international, national, regional and local), where different scales do not compete, but rather cooperate

7.3.5 Future challenges

In order to achieve this impact and to realize the promising opportunities, several critical challenges will need to be overcome.

Up to now, rigorous theories that describe HBMs are still lacking. Virtually no understanding exists about how specific features at distinct scales interact, and for example participate in deformation. However, such models are vital to arrive at a solution for the universality-diversity question and Nature's hierarchical material design. The path to success is to develop cross-scale relationships and constitutive equations for different hierarchical scales within the structure-property paradigm of materials science (see Figure 7.2), that is, to understand if and how nano-/meso-/micro-changes affect properties at larger scales. To achieve this goal, structural architecture will have to be considered across the scales, possibly combined with fractal theory [257], and investigated in light of the UDP. By developing the Hierarchical Bell Model, we did the first step towards this direction. This theory allows us to cover for the first time effects on different scales simultaneously. For instance, effects from point mutations, which do not change the secondary structure of AHs, but which lead to stutter defects can now be covered. However, more effort in developing this kind of theories will be necessary during the next years.

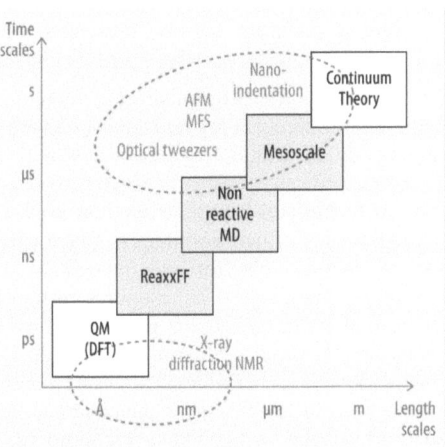

Figure 7.6: Schematic that illustrates the concept of hierarchical multi-scale modeling (schematic adapted from [14]). Hierarchical coupling of different computational tools can be used to traverse throughout a wide range of length- and time scales. Such methods enable to provide a fundamental insight into deformation and fracture phenomena, across various time and length scales. Handshaking between different methods enables one to transport information from one scale to another. Eventually, results of atomistic, molecular or mesoscale simulation may feed into constitutive equations or continuum models. While continuum mechanical theories have been very successful for crystalline materials, PMs require statistical theories, e.g. the Hierarchical Bell Model [24, 26]. Experimental techniques such as Atomic Force Microscope (AFM), Molecular Force Spectroscopy (MFS), nanoindentation or optical tweezers now overlap into atomistic and molecular approaches, enabling direct comparison of experiment and simulation [35].

Furthermore, an appropriate nomenclature to describe, characterize and analyze HBMs is still missing. Definitions and measures for material properties such as hierarchical degree, level of robustness, degree of universality, and others are crucial, and the terminology for cross scale relations such as scale separation, integration and interaction must be defined. We hope that the work presented here will stimulate extensive research in this direction.

Computational modeling techniques have progressed enormously during the last few years (see e.g. Figure 3.3), and simulation techniques like MD find broad application and increasing acceptance. But these simulation approaches are still limited to samples of a few nanometers in size and modeling techniques, linking atomistic to continuum scale in biological materials, which lack a regular atomic lattice, are in a very early stage of development. To overcome these limits new numerical models, e.g. universal platforms combining different scripts, and new computational architectures will be necessary, followed by new data analysis and visualization tools.

In addition to the computational techniques, experimental techniques at the level of individual molecules progressed immensely during the last decade (see, e.g. Figure 7.6 for an overlay of simulation techniques with experimental methods such as AFM, optical tweezers and others [35]). It is vital to quantify how much the resolution and stability of current techniques can increase and how good new technologies will be able to resolve these issues. Further techniques, which allow detailed research on and the control of the intermolecular behavior are necessary. For example the highly complex process of assembly, future key in manufacturing these materials need to be understood and controlled in its whole in order to allow future applications. In other words, beyond experimental issues, manufacturing challenges need to be overcome, such as the application of recombinant DNA techniques for industrial volumes of material production, or the design of macro-materials from nano-devices. Will it ever be possible to produce and utilize those technologies and materials that researchers are currently investigating? Will we ever be able to merge the process of material synthesis and its application? And if yes, which risks and benefits will these new technologies entail for society?

Even if the new field of biomimetics on the nanoscale has been a field with high potential, it had few triumphs so far. One of the few known and broadly applied examples is the lotus effect. However many effects appearing on this scale such as the gecko principle of adhesion has been understood and prototypes of tape have been developed but not yet commercialized [258]. This example shows that most principles understood thus far are of possible future value rather than current successes. In order to make commercial applications from scientifically understood principles and thus reach the steep part of the bio-nano-technology S-curve will require engineers to work more extensively together with nano-scientists in the next decade.

Another important aspect of the work is the impact in the medical sciences. Up to now the main focus was on molecular and genetic aspects on this level. The recently introduced material aspect could facilitate the understanding of the mechanistic aspects of genetic diseases, the effect of diseases such as cancer on the behavior of protein networks (e.g. lamin protein structures), which could lead to new breakthroughs for diagnostics and disease treatment options. Similarly to the commercialization of nano science, we have to continue the investigations made up to now and hope that the visions of today will become reality in the future.

8 Outlook

Overcoming the challenges listed in Section 7.3.5 may require a convergence of disciplines in two regards: First, experimental, theoretical and computational approaches must be combined extensively, in order to understand, explain and successfully apply observed phenomena present in the biological nano-world (e.g. Figure 7.5 and Figure 7.6). Second, scientists from different disciplines such as Physics, Chemistry, Biology, Engineering, Computer Science and Medicine must work together. All these fields are crucial for the understanding of the biological nano-science and the future application of the newly generated knowledge for novel technologies. It is even conceivable that there might be a transition from a multi-disciplinary approach to the creation of a new discipline and scientific organizations. In this Thesis basic physical and biological science was combined with an engineering perspective on materials design.

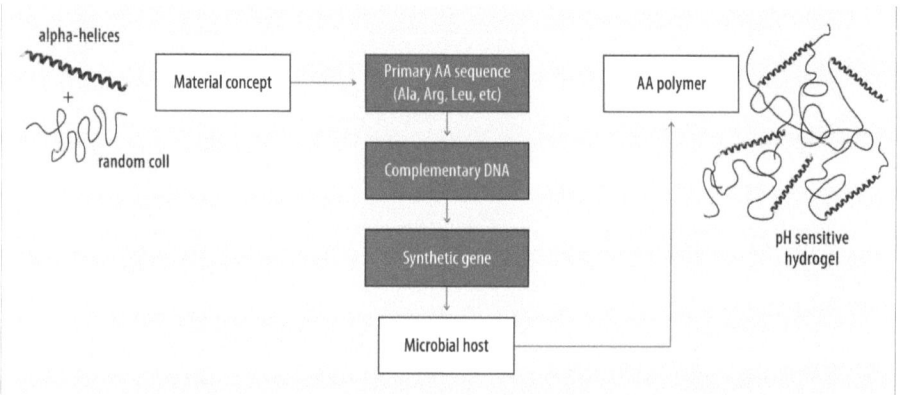

Figure 7.7: Recombinant DNA technique, as pioneered by Tirrell and others [1-4] may be a viable strategy to synthesize novel protein based materials. These methods enable one to combine distinct structural protein motifs from different sources into new materials, via formation of a proper DNA, a synthetic gene, and the production of these proteins in microbial hosts, eventually leading to the AA polymer. These techniques provide ultimate control over the primary structure of proteins, and can thereby used to define larger, hierarchical structural features. The use of artificial AA beyond the 20 naturally occurring ones opens additional opportunities.

Applications of the new materials and structures that result from these studies are new biomaterials, new polymers, new composites, engineered spider silk, new scaffolding tissues, improved understanding of cell-extracellular matrix material interactions, cell mechanics, hierarchical structures and self-assembly.

Recombinant DNA technique, as pioneered by Tirrell *et al.* [1-4], may be a viable strategy to synthesize novel protein based materials (see Figure 7.7). These methods enable one to combine distinct structural protein motifs from different sources into new materials, via formation of a proper DNA, a synthetic gene, and the production of these proteins in microbial hosts, eventually leading to the AA polymer. These techniques provide ultimate control over the primary structure of proteins, and can thereby used to define larger, hierarchical structural features. The use of artificial AAs beyond the 20 naturally occurring ones opens additional opportunities.

In addition to the long-term impact in biology, bioengineering and medicine, this research may eventually contribute to our theoretical understanding of how structural features at different scales interact with one another. In light of the 'extended physical design space' discussed in this Thesis, this may transform engineering approaches not only for materials applications, but also in manufacturing, transportation or designs of networks.

APPENDIX

9 List of abbreviations and important mathematical symbols

Abbreviation	Description
AA	Amino acid
AH	Alpha-helix, alpha-helical
AFM	Atomic force microscopy
AP	Angular point (when HB rupture sets in)
AR	Asymptotic regime
HB	Hydrogen bond
HBMs, PMs	Hierarchical biological materials, protein materials
BS	Beta-sheet
CC	Coiled-coil
FDM	Fast deformation mode (sequential HB rupture)
IF	Intermediate filament
MD	Molecular dynamics
MF	Microfilaments, actin filaments
MT	Microtubule
PDB	Protein Data Bank
RMSD	Root mean square distance
SDM	Slow deformation mode (simultaneous rupture of HBs)
SMD	Steered molecular dynamics
UDP	Universality-diversity paradigm
VMD	Visual molecular dynamics
WLC	Worm-like-chain model

Mathematical symbol	Unit	Description
α	---	Ration of end to end length and contour length
α_{cr}	---	α at which rupture sets in
α_c	° (degrees)	Angle between the direction of Coulomb bond and the direction of applied load
b_i	---	Number of parallel elements on hierarchy i
E_b	kcal/mol	High of the effective energy barrier at the transition state
E_b^0	kcal/mol	Energy barrier of one HB
ε	%	Engineering strain, displacement normalized by the length

f	pN		Force per bond
F	pN		Force applied to the whole system
f_b	pN		Thermal force $k_b \cdot T / x_b$
F_{cr}, v_{cr}	pN, m/s		Force and velocity, which separate the FDM from the SDM
F_{hi}	pN		Force applied to a system consisting of i hierarchies
f_{AR}, f_{SDM}, f_{FDM}	pN		Force in the AR, SDM and FDM
F_{tens}	pN		Tensile force leading to protein unfolding
F_{shear}, f_{shear}	pN		Shear force, leading to sliding between proteins (overall and per bond)
F_v	pN		Force contribution from pulling speed dependence
G	kcal/mol		Free energy
γ	kcal/mol/Å		Energy stored per unit length of an AH
h_o	---		Basis hierarchy describing individual HBs
h_i	---		Hierarchical level i of a system
K_0	kcal/mol/Å²		Spring constant of a pulling cantilever
k_B	J/K		Boltzmann constant
k_c	---		Number of Coulomb bonds, at which shear will set in
k_i	---		Number of elements on hierarchy i breaking simultaneously
k_{SMD}	kcal/mol/Å²		Spring constant for SMD simulations
l_i	---		Number of serial elements on hierarchical level i
n_c	---		Overall number of Coulomb bonds, being present between molecules
q_i	e		Charge of residue i (in elementary charges)
r	%		Robustness, ratio of strength of a failed system and an intact system (0%-100%)
σ	MPa		Engineering stress
θ	° (degrees)		Angle between pulling direction and reaction coordinate
τ_{eq}	s		Time scale at which a system is in equilibrium
T	K		Absolute temperature
v	m/s		Macroscopically: pulling speed, microscopically bond breaking speed
v_0, v_{eq}	m/s		Natural bond breaking speed, when no load is applied or when system is in equilibrium
ω_0	s⁻¹		Natural bond vibration frequency (fixed value, 10^{-13} s⁻¹)
χ	s⁻¹		Off rate: bond dissociation per second, $1/\tau$
x_b	Å		Location of the energy barrier referred to the equilibrium
ξ_P	Å		Persistence length

10 References

1. Langer, R. and D.A. Tirrell, *Designing materials for biology and medicine.* Nature, 2004. **428**(6982): p. 487-492.
2. Petka, W.A., et al., *Reversible hydrogels from self-assembling artificial proteins.* Science, 1998. **281**(5375): p. 389-392.
3. Michon, T. and D.A. Tirrell, *Artificial proteins.* Biofutur, 2000. **2000**(197): p. 34-38.
4. van Hest, J.C.M. and D.A. Tirrell, *Protein-based materials, toward a new level of structural control.* Chemical Communications, 2001(19): p. 1897-1904.
5. Buehler, M.J., *Atomistic and continuum modeling of mechanical properties of collagen: Elasticity, fracture and self-assembly.* J. Mater. Res., 2006. **21**(8): p. 1947-1961.
6. Fratzl, P., et al., *Structure and mechanical quality of the collagen-mineral nano-composite in bone.* Journal Of Materials Chemistry, 2004. **14**(14): p. 2115-2123.
7. An, K.N., Y.L. Sun, and Z.P. Luo, *Flexibility of type I collagen and mechanical property of connective tissue.* Biorheology, 2004. **41**(3-4): p. 239-246.
8. Ramachandran, G.N., Kartha, G., *Structure of collagen.* Nature, 1955. **176**: p. 593–595.
9. Toivola, D.M., et al., *Cellular integrity plus: organelle-related and protein-targeting functions of intermediate filaments.* Trends in Cell Biology, 2005. **15**(11): p. 608-617.
10. Kim, S. and P.A. Coulombe, *Intermediate filament scaffolds fulfill mechanical, organizational, and signaling functions in the cytoplasm.* Genes & Development, 2007. **21**(13): p. 1581-1597.
11. Wagner, O.I., et al., *Softness, strength and self-repair in intermediate filament networks.* Experimental Cell Research, 2007. **313**(10): p. 2228-2235.
12. Buehler, M.J., *Atomistic modeling of materials failure.* 2008: Springer (New York).
13. Ackbarow, T., Buehler M.J., , *Molecular mechanics of stutter defects in vimentin intermediate filaments* available online: DOI 10.1007/s11340-007-9100-6, 2008.
14. Buehler, M.J. and T. Ackbarow, *Fracture mechanics of protein materials.* Materials Today, 2007. **10**(9): p. 46-58.
15. Alberts, B., et al., *Molecular Biology of the Cell.* 2002: Taylor & Francis.
16. Pareto, V., *Manual of Political Economy.* 1909, New York.
17. Chen, Y.S., P.P. Chong, and M.Y. Tong, *Mathematical and computational modeling of the pareto principle.* Mathematical and Computer Modelling, 1994. **19**(9): p. 61-80.
18. Chen, Y.S., P.P. Chong, and Y.G. Tong, *Theoretical foundation of the 80/20 rule.* Scientometrics, 1993. **28**(2): p. 183-204.
19. Fratzl, P., et al., *Structure and mechanical quality of the collagen–mineral nano-composite in bone.* Journal of Materials Chemistry, 2004. **14**(14): p. 2115-2123.
20. Bell, G.I., *Models for Specific Adhesion of Cells to Cells.* Science, 1978. **200**(4342): p. 618-627.
21. Zhang, H., T. Ackbarow, and M.J. Buehler, *Muscle dystrophy single point mutation in the 2B segment of lamin A does not affect the mechanical properties at the dimer level.* Journal of Biomechanics, 2008. **41**(6): p. 1295-1301.

22. Fudge, D.S., et al., *The mechanical properties of hydrated intermediate filaments: Insights from hagfish slime threads.* Biophysical Journal, 2003. **85**(3): p. 2015-2027.
23. Kreplak, L., U. Aebi, and H. Herrmann, *Molecular mechanisms underlying the assembly of intermediate filaments.* Experimental Cell Research, 2004. **301**(1): p. 77-83.
24. Ackbarow, T. and M.J. Buehler, *Superelasticity, energy dissipation and strain hardening of vimentin coiled-coil intermediate filaments: atomistic and continuum studies.* Journal of Materials Science, 2007. **42**(21): p. 8771-8787.
25. Sheu, S.-Y., et al., *Energetics of hydrogen bonds in peptides.* PNAS, 2003. **100**(22): p. 12683-12687.
26. Ackbarow, T., et al., *Hierarchies, multiple energy barriers, and robustness govern the fracture mechanics of alpha-helical and beta-sheet protein domains.* Proceedings of the National Academy of Sciences of the United States of America, 2007. **104**: p. 16410-16415.
27. Kramers, H.A., *Brownian motion in a field of force and the diffusion model of chemical reactions.* Physica, 1940. **7**: p. 10.
28. Aebi, U., et al., *The Nuclear Lamina is a Meshwork of intermediate-type filaments.* Nature, 1986. **323**(6088): p. 560-564.
29. Strelkov, S.V., H. Herrmann, and U. Aebi, *Molecular architecture of intermediate filaments.* Bioessays, 2003. **25**(3): p. 243-251.
30. Ashby, M.F., et al., *The Mechanical Properties of Natural Materials. I. Material Property Charts.* Proceedings: Mathematical and Physical Sciences, 1995. **450**(1938): p. 123-140.
31. Smith, T.A., et al., *Modeling alpha-helical coiled-coil interactions: The axial and azimuthal alignment of 1B segments from vimentin intermediate filaments.* Proteins-Structure Function and Genetics, 2003. **50**(2): p. 207-212.
32. Herrmann, H. and U. Aebi, *Intermediate filaments: Molecular structure, assembly mechanism, and integration into functionally distinct intracellular scaffolds.* Annual Review of Biochemistry, 2004. **73**: p. 749-789.
33. Mucke, N., et al., *Assessing the flexibility of intermediate filaments by atomic force microscopy.* Journal of Molecular Biology, 2004. **335**(5): p. 1241-1250.
34. Strelkov, S.V., et al., *Crystal structure of the human lamin a coil 2B dimer: Implications for the head-to-tail association of nuclear lamins.* Journal of Molecular Biology, 2004. **343**(4): p. 1067-1080.
35. Lim, C.T., et al., *Experimental techniques for single cell and single molecule biomechanics.* Materials Science & Engineering C-Biomimetic and Supramolecular Systems, 2006. **26**(8): p. 1278-1288.
36. Wang, N. and D. Stamenovic, *Mechanics of vimentin intermediate filaments.* Journal of Muscle Research and Cell Motility, 2002. **23**(5-6): p. 535-540.
37. Wang, T., et al., *Nanorheology measurement on single circularly permuted green fluorescent protein molecule.* Colloids and Surfaces B-Biointerfaces, 2005. **40**(3-4): p. 183-187.
38. Buehler, M.J., *Hierarchical chemo-nanomechanics of proteins: Entropic elasticity, protein unfolding and molecular fracture.* Journal of Mechanics of Materials and Structures, 2007. **2**(6): p. 1019-1057.
39. Seifert, U., *Rupture of multiple parallel molecular bonds under dynamic loading.* Physical Review Letters, 2000. **84**(12): p. 2750-2753.

40. Erdmann, T. and U.S. Schwarz, *Stability of adhesion clusters under constant force.* Phys Rev Lett, 2004. **92**(10): p. 108102.
41. Burkhard, P., et al., *The coiled-coil trigger site of the rod domain of cortexillin I unveils a distinct network of interhelical and intrahelical salt bridges.* Structure with Folding & Design, 2000. **8**(3): p. 223-230.
42. Janmey, P.A., et al., *Viscoelastic Properties of Vimentin Compared with Other Filamentous Biopolymer Networks.* Journal of Cell Biology, 1991. **113**(1): p. 155-160.
43. Dietz, H. and M. Rief, *Elastic bond network model for protein unfolding mechanics.* Physical Review Letters, 2008. **1**(9): p. 4.
44. Buehler, M.J., *Nano- and micromechanical properties of hierarchical biological materials and tissues* Journal of Materials Science, 2007 **42**(21): p. 8765-8770.
45. Fratzl, P. and R. Weinkamer, *Nature's hierarchical materials.* Progress in Materials Science, 2007. **52**(8): p. 1263-1334.
46. R. Ritchie, M.J.B., P. Hansma, *Plasticity and toughness in bone: A problem of multiple length-scales.* Physics Today, in press.
47. Buehler, M.J., *Nature designs tough collagen: Explaining the nanostructure of collagen fibrils.* Proceedings of the National Academy of Sciences of the United States of America, 2006. **103**(33): p. 12285-12290.
48. Rose, A. and I. Meier, *Scaffolds, levers, rods and springs: diverse cellular functions of long coiled-coil proteins.* Cellular and Molecular Life Sciences, 2004. **61**(16): p. 1996-2009.
49. Maccallum, J.L., et al., *Hydrophobic association of {alpha}-helices, steric dewetting, and enthalpic barriers to protein folding.* Proc Natl Acad Sci U S A, 2007. **104**(15): p. 6206-10.
50. Gruber, M. and A.N. Lupas, *Historical review: Another 50th anniversary - new periodicities in coiled coils.* Trends in Biochemical Sciences, 2003. **28**(12): p. 679-685.
51. Mucke, N., et al., *Molecular and biophysical characterization of assembly-starter units of human vimentin.* Journal of Molecular Biology, 2004. **340**(1): p. 97-114.
52. McLachlan, A.D. and J. Karn, *Periodic Features in the Amino-Acid-Sequence of Nematode Myosin Rod.* Journal of Molecular Biology, 1983. **164**(4): p. 605-626.
53. Brown, J.H., C. Cohen, and D.A.D. Parry, *Heptad breaks in alpha-helical coiled coils: Stutters and stammers.* Proteins-Structure Function and Genetics, 1996. **26**(2): p. 134-145.
54. Strelkov, S.V., et al., *Conserved segments 1A and 2B of the intermediate filament dimer: their atomic structures and role in filament assembly.* EMBO Journal, 2002. **21**(6): p. 1255-1266.
55. Herrmann, H. and U. Aebi, *Intermediate filament assembly: temperature sensitivity and polymorphism.* Cellular And Molecular Life Sciences, 1999. **55**(11): p. 1416-1431.
56. Parry, D.A.D., *Fibrinogen - Preliminary-Analysis of Amino-Acid Sequences of Portions of Alpha-Chains, Beta-Chains and Gamma-Chains Postulated to Form Interdomainal Link between Globular Regions of Molecule.* Journal of Molecular Biology, 1978. **120**(4): p. 545-551.
57. Helfand, B.T., L. Chang, and R.D. Goldman, *Intermediate filaments are dynamic and motile elements of cellular architecture.* Journal of Cell Science, 2004. **117**(2): p. 133-141.

58. Fudge, D.S. and J.M. Gosline, *Molecular design of the alpha-keratin composite: insights from a matrix-free model, hagfish slime threads.* Proceedings of the Royal Society of London Series B-Biological Sciences, 2004. **271**(1536): p. 291-299.
59. Moir, R.D. and T.P. Spann, *The structure and function of nuclear lamins: implications for disease.* Cellular and Molecular Life Sciences, 2001. **58**(12-13): p. 1748-1757.
60. Wilson, K.L., M.S. Zastrow, and K.K. Lee, *Lamins and disease: Insights into nuclear infrastructure.* Cell, 2001. **104**(5): p. 647-650.
61. Ingber, D.E., *Cellular mechanotransduction: putting all the pieces together again.* Faseb Journal, 2006. **20**(7): p. 811-827.
62. Ingber, D.E., et al., *Cellular tensegrity - exploring how mechanical changes in the cytoskeleton regulate cell-growth, migration, and tissue pattern during morphogenesis.* International Review of Cytology - a Survey of Cell Biology, Vol 150, 1994. **150**: p. 173-224.
63. Coulombe, P.A., et al., *The 'ins' and 'outs' of intermediate filament organization.* Trends in Cell Biology, 2000. **10**(10): p. 420-428.
64. Kreplak, L., et al., *Exploring the mechanical behavior of single intermediate filaments.* Journal of Molecular Biology, 2005. **354**(3): p. 569-577.
65. Kreplak, L., H. Herrmann, and U. Aebi, *Tensile Properties of Single Desmin Intermediate Filaments.* Biophys. J., 2008. **94**(7): p. 2790-2799.
66. Kiss, B., A. Karsai, and M.S.Z. Kellermayer, *Nanomechanical properties of desmin intermediate filaments.* Journal of Structural Biology, 2006. **155**(2): p. 327-339.
67. Omary, M.B., P.A. Coulombe, and W.H.I. McLean, *Mechanisms of disease: Intermediate filament proteins and their associated diseases.* New England Journal of Medicine, 2004. **351**(20): p. 2087-2100.
68. Schietke, R., et al., *Mutations in vimentin disrupt the cytoskeleton in fibroblasts and delay execution of apoptosis.* European Journal of Cell Biology, 2006. **85**(1): p. 1-10.
69. Bridger, J.M., et al., *The nuclear lamina - Both a structural framework and a platform for genome organization.* Febs Journal, 2007. **274**(6): p. 1354-1361.
70. Bridger, J.M., et al., *The nuclear lamina. Both a structural framework and a platform for genome organization.* Febs J, 2007. **274**(6): p. 1354-61.
71. Dahl, K.N., et al., *Distinct structural and mechanical properties of the nuclear lamina in Hutchinson-Gilford progeria syndrome.* Proceedings of the National Academy of Sciences of the United States of America, 2006. **103**(27): p. 10271-10276.
72. Broers, J.L., C.J. Hutchison, and F.C. Ramaekers, *Laminopathies.* J Pathol, 2004. **204**(4): p. 478-88.
73. Prokocimer, M., A. Margalit, and Y. Gruenbaum, *The nuclear lamina and its proposed roles in tumorigenesis: Projection on the hematologic malignancies and future targeted therapy.* Journal of Structural Biology, 2006. **155**(2): p. 351-360.
74. Dahl, K.N., et al., *The nuclear envelope lamina network has elasticity and a compressibility limit suggestive of a molecular shock absorber.* Journal of Cell Science, 2004. **117**(20): p. 4779-4786.
75. Lammerding, J., et al., *Lamin A/C deficiency causes defective nuclear mechanics and mechanotransduction.* Journal of Clinical Investigation, 2004. **113**(3): p. 370-378.

76. Sullivan, T., et al., *Loss of A-type lamin expression compromises nuclear envelope integrity leading to muscular dystrophy.* Journal of Cell Biology, 1999. **147**(5): p. 913-919.
77. Houben, F., et al., *Role of nuclear lamina-cytoskeleton interactions in the maintenance of cellular strength.* Biochim Biophys Acta, 2007. **1773**(5): p. 675-86.
78. Brown, C.A., et al., *Novel and recurrent mutations in lamin A/C in patients with Emery-Dreifuss muscular dystrophy.* American Journal of Medical Genetics, 2001. **102**(4): p. 359-367.
79. Fidzianska, A., D. Toniolo, and I. Hausmanowa-Petrusewicz, *Ultrastructural abnormality of sarcolemmal nuclei in Emery-Dreifuss muscular dystrophy (EDMD).* Journal of the Neurological Sciences, 1998. **159**(1): p. 88-93.
80. Vashishta, P., R.K. Kalia, and A. Nakano, *Large-scale atomistic simulations of dynamic fracture.* Comp. in Science and Engrg., 1999: p. 56-65.
81. Rountree, C.L., et al., *Atomistic aspects of crack propagation in brittle materials: Multimillion atom molecular dynamics simulations.* Annual Rev. of Materials Research, 2002. **32**: p. 377-400.
82. Buehler, M.J. and H.J. Gao, *Dynamical fracture instabilities due to local hyperelasticity at crack tips.* Nature, 2006. **439**(7074): p. 307-310.
83. Buehler, M.J., F.F. Abraham, and H. Gao, *Hyperelasticity governs dynamic fracture at a critical length scale.* Nature, 2003. **426**: p. 141-146.
84. Buehler, M.J. and H. Gao, *Ultra large scale atomistic simulations of dynamic fracture* in *Handbook of Theoretical and Computational Nanotechnology, Edited by W. Schommers and A. Rieth.* 2006, American Scientific Publishers (ASP).
85. Buehler, M.J., A.C.T.v. Duin, and W.A. Goddard, *Multi-paradigm modeling of dynamical crack propagation in silicon using the ReaxFF reactive force field.* Phys. Rev. Lett., 2006. **96**(9): p. 095505
86. Buehler, M.J., et al., *Threshold Crack Speed Controls Dynamical Fracture of Silicon Single Crystals.* Phys. Rev. Lett., 2007. **99**: p. 165502
87. Wang, W., et al., *Biomolecular simulations: Recent developments in force fields, simulations of enzyme catalysis, protein-ligand, protein-protein, and protein-nucleic acid noncovalent interactions.* Annual Review Of Biophysics And Biomolecular Structure, 2001. **30**: p. 211-243.
88. Mackerell, A.D., *Empirical force fields for biological macromolecules: Overview and issues.* Journal Of Computational Chemistry, 2004. **25**(13): p. 1584-1604.
89. Phillips, J.C., et al., *Scalable molecular dynamics with NAMD.* Journal Of Computational Chemistry, 2005. **26**(16): p. 1781-1802.
90. Nelson, M.T., et al., *NAMD: A parallel, object oriented molecular dynamics program.* International Journal Of Supercomputer Applications And High Performance Computing, 1996. **10**(4): p. 251-268.
91. MacKerell, A.D., et al., *All-atom empirical potential for molecular modeling and dynamics studies of proteins.* Journal Of Physical Chemistry B, 1998. **102**(18): p. 3586-3616.
92. Anderson, D., *Collagen Self-Assembly: A Complementary Experimental and Theoretical Perspective.* 2005, University of Toronto: Toronto, Canada.
93. Mayo, S.L., B.D. Olafson, and W.A. Goddard, *Dreiding - A Generic Force-Field For Molecular Simulations.* Journal Of Physical Chemistry, 1990. **94**(26): p. 8897-8909.

94. Rappe, A.K., et al., *Uff, A Full Periodic-Table Force-Field For Molecular Mechanics And Molecular-Dynamics Simulations.* Journal Of The American Chemical Society, 1992. **114**(25): p. 10024-10035.
95. Pearlman, D.A., et al., *Amber, A Package Of Computer-Programs For Applying Molecular Mechanics, Normal-Mode Analysis, Molecular-Dynamics And Free-Energy Calculations To Simulate The Structural And Energetic Properties Of Molecules.* Computer Physics Communications, 1995. **91**(1-3): p. 1-41.
96. Goddard, W.A., *A Perspective of Materials Modeling* in *Handbook of Materials Modeling*, S. Yip, Editor. 2006, Springer.
97. Bernstein, F.C., et al., *Protein data bank - computer-based archival file for macromolecular structures.* European Journal of Biochemistry, 1977. **80**(2): p. 319-324.
98. Lu, H., et al., *Unfolding of titin immunoglobulin domains by steered molecular dynamics simulation.* Biophysical Journal, 1998. **75**(2): p. 662-671.
99. Buehler, M.J.a.H.G., *Ultra large scale atomistic simulations of dynamic fracture*, in *Handbook of Theoretical and Computational Nanotechnology*, W.S.a.A. Rieth, Editor. 2006, American Scientific Publishers (ASP).
100. Kadau, K., T.C. Germann, and P.S. Lomdahl, *Large-Scale Molecular-Dynamics Simulation of 19 Billion particles.* Int. J. Mod. Phys. C, 2004. **15**: p. 193.
101. Humphrey, W., A. Dalke, and K. Schulten, *VMD: Visual molecular dynamics.* Journal Of Molecular Graphics, 1996. **14**(1): p. 33.
102. Tsai, D.H., *Virial theorem and stress calculation in molecular-dynamics.* J. of Chemical Physics, 1979. **70**(3): p. 1375-1382.
103. Zimmerman, J.A., et al., *Calculation of stress in atomistic simulation.* Model. Sim. Mat. Science and Engr., 2004. **12**: p. S319-S332.
104. Bustamante, C., et al., *Entropic Elasticity Of Lambda-Phage Dna.* Science, 1994. **265**(5178): p. 1599-1600.
105. Marko, J.F. and E.D. Siggia, *Stretching DNA.* Macromolecules, 1995. **28**(26): p. 8759-8770.
106. Prater, C.B., H.J. Butt, and P.K. Hansma, *Atomic force microscopy.* Nature, 1990. **345**(6278): p. 839-840.
107. Bozec, L., et al., *Atomic force microscopy of collagen structure in bone and dentine revealed by osteoclastic resorption.* Ultramicroscopy, 2005. **105**(1-4): p. 79-89.
108. Guzman, C., et al., *Exploring the mechanical properties of single vimentin intermediate filaments by atomic force microscopy.* Journal of Molecular Biology, 2006. **360**(3): p. 623-630.
109. Yuan, C.B., et al., *Energy landscape of streptavidin-biotin complexes measured by atomic force microscopy.* Biochemistry, 2000. **39**(33): p. 10219-10223.
110. Sun, Y.L., et al., *Stretching type II collagen with optical tweezers.* Journal Of Biomechanics, 2004. **37**(11): p. 1665-1669.
111. Dao, M., C.T. Lim, and S. Suresh, *Mechanics of the human red blood cell deformed by optical tweezers.* Journal Of The Mechanics And Physics Of Solids, 2003. **51**(11-12): p. 2259-2280.
112. Sun, Y.L., Z.P. Luo, and K.N. An, *Stretching short biopolymers using optical tweezers.* Biochemical And Biophysical Research Communications, 2001. **286**(4): p. 826-830.

113. Tozzini, V., *Coarse-grained models for proteins*. Current Opinion in Structural Biology, 2005. **15**(2): p. 144-150.
114. Southern, J., et al., *Multi-scale computational modelling in biology and physiology*. Progress in Biophysics & Molecular Biology, 2008. **96**(1-3): p. 60-89.
115. Yaliraki, S.N. and M. Barahona, *Chemistry across scales: from molecules to cells*. Philosophical Transactions of the Royal Society a-Mathematical Physical and Engineering Sciences, 2007. **365**(1861): p. 2921-2934.
116. Broberg, K.B., *Cracks and Fracture*. 1990: Academic Press.
117. Freund, L.B., *Dynamic Fracture Mechanics*. 1990: Cambridge University Press, ISBN 0-521-30330-3.
118. Buehler, M.J., *Large-scale hierarchical molecular modeling of nano-structured biological materials*. Journal of Computational and Theoretical Nanoscience, 2006. **3**(5): p. 603–623.
119. Dietz, H., et al., *Anisotropic deformation response of single protein molecules*. Proceedings of the National Academy of Sciences of the United States of America, 2006. **103**(34): p. 12724-12728.
120. Knowles, T.P., et al., *Role of intermolecular forces in defining material properties of protein nanofibrils*. Science, 2007. **318**(5858): p. 1900-1903.
121. Hirth, J.P. and J. Lothe, *Theory of Dislocations*. 1982: Wiley-Interscience.
122. Lakes, R., *Materials with Structural Hierarchy*. Nature, 1993. **361**(6412): p. 511-515.
123. Ahl, V., Allen, T.F.H, *Hierarchy Theory - A Vision, Vocabulary, and Epistemology*. 1996, New York: Columbia University Press.
124. Hanggi, P., P. Talkner, and M. Borkovec, *Reaction-rate theory: fifty years after Kramers*. Rev. Mod. Phys, 1990. **62**(2): p. 251–341.
125. Zhurkov, S.N., *Kinetic concept of the strength of solids*. Int. Journal of Fracture Mechanics, 1965. **1**: p. 311-323.
126. Evans, E.A. and D.A. Calderwood, *Forces and Bond Dynamics in Cell Adhesion*. Science, 2007. **316**(5828): p. 1148-1153.
127. Bayas, M.V., et al., *Lifetime measurements reveal kinetic differences between homophilic cadherin bonds*. Biophysical Journal, 2006. **90**(4): p. 1385-1395.
128. Evans, E., *Mechanical switching and cooperative coupling of unbinding pathways in bioadhesion bonds*. Abstracts of Papers of the American Chemical Society, 2004. **227**: p. U469-U470.
129. Evans, E., *Probing the relation between force - Lifetime - and chemistry in single molecular bonds*. Annual Review of Biophysics and Biomolecular Structure, 2001. **30**: p. 105-128.
130. Evans, E., et al., *Chemically distinct transition states govern rapid dissociation of single L-selectin bonds under force*. Proceedings of the National Academy of Sciences of the United States of America, 2001. **98**(7): p. 3784-3789.
131. Evans, E.B., *Looking inside molecular bonds at biological interfaces with dynamic force spectroscopy*. Biophysical Chemistry, 1999. **82**(2-3): p. 83-97.
132. Merkel, R., et al., *Energy landscapes of receptor-ligand bonds explored with dynamic force spectroscopy*. Nature (London), 1999. **379**(6714): p. 50-53.
133. Evans, E. and K. Ritchie, *Dynamic strength of molecular adhesion bonds*. Biophysical Journal, 1997. **72**(4): p. 1541-1555.

134. Dudko, O.K., G. Hummer, and A. Szabo, *Intrinsic rates and activation free energies from single-molecule pulling experiments.* Physical Review Letters, 2006. **96**(10).
135. Dudko, O.K., et al., *Extracting Kinetics from Single-Molecule Force Spectroscopy: Nanopore Unzipping of DNA Hairpins.* Biophys. J., 2007. **92**(12): p. 4188-4195.
136. Hummer, G. and A. Szabo, *Kinetics from nonequilibrium single-molecule pulling experiments.* Biophysical Journal, 2003. **85**(1): p. 5-15.
137. Hummer, G. and A. Szabo, *Free energy surfaces from single-molecule force spectroscopy.* Accounts of Chemical Research, 2005. **38**(7): p. 504-513.
138. Hummer, G. and A. Szabo, *Free energy reconstruction from nonequilibrium single-molecule pulling experiments.* Proceedings of the National Academy of Sciences of the United States of America, 2001. **98**(7): p. 3658-3661.
139. Jarzynski, C., *Equilibrium free-energy differences from nonequilibrium measurements: A master-equation approach.* Physical Review E, 1997. **56**(5): p. 5018-5035.
140. Jarzynski, C., *Nonequilibrium equality for free energy differences.* Physical Review Letters, 1997. **78**(14): p. 2690-2693.
141. Makarov, D.E., *Unraveling Individual Molecules by Mechanical Forces: Theory Meets Experiment.* Biophys. J., 2007. **92**(12): p. 4135-4136.
142. Li, P.C. and D.E. Makarov, *Theoretical studies of the mechanical unfolding of the muscle protein titin: Bridging the time-scale gap between simulation and experiment.* Journal of Chemical Physics, 2003. **119**(17): p. 9260-9268.
143. Schlierf, M. and M. Rief, *Single-Molecule Unfolding Force Distributions Reveal a Funnel-Shaped Energy Landscape.* Biophys. J., 2006. **90**(4): p. L33-35.
144. Gilli, P., et al., *Covalent versus electrostatic nature of the strong hydrogen bond: Discrimination among single, double, and asymmetric single-well hydrogen bonds by variable-temperature X-ray crystallographic methods in beta-diketone enol RAHB systems.* Journal of the American Chemical Society, 2004. **126**(12): p. 3845-3855.
145. Wiita, A.P., et al., *Force-dependent chemical kinetics of disulfide bond reduction observed with single-molecule techniques.* Proceedings of the National Academy of Sciences of the United States of America, 2006. **103**(19): p. 7222-7227.
146. Maloney, C.E. and D.J. Lacks, *Energy barrier scalings in driven systems.* Physical Review E, 2006. **73**(6).
147. West, D.K., P.D. Olmsted, and E. Paci, *Mechanical unfolding revisited through a simple but realistic model.* Journal of Chemical Physics, 2006. **124**(15).
148. Bell, G.I., *Models for the specific adhesion of cells to cells.* Science, 1978. **200**(4342): p. 618-627.
149. Erdmann, T. and U.S. Schwarz, *Stability of Adhesion Clusters under Constant Force.* Physical Review Letters, 2004. **92**(10): p. 108102.
150. Erdmann, T. and U.S. Schwarz, *Bistability of Cell-Matrix Adhesions Resulting from Nonlinear Receptor-Ligand Dynamics.* Biophysical Journal, 2006. **91**(6): p. L60.
151. Keten, S. and M.J. Buehler, *Geometric confinement governs the rupture strength of H-bond assemblies at a critical length scale.* Nano Letters, 2008. **8**: p. 743-748.
152. Keten, S. and M.J. Buehler, *Asymptotic strength limit of hydrogen-bond assemblies in proteins at vanishing pulling rates.* Physical Review Letters, 2008. **100**(19).

153. Lammerding, J., et al., *Lamins A and C but not lamin B1 regulate nuclear mechanics.* Journal of Biological Chemistry, 2006. **281**(35): p. 25768-25780.
154. Bonne, G., et al., *Clinical and molecular genetic spectrum of autosomal dominant Emery-Dreifuss muscular dystrophy due to mutations of the lamin A/C gene.* Annals of Neurology, 2000. **48**(2): p. 170-180.
155. McManus, J.J., et al., *From the Cover: Altered phase diagram due to a single point mutation in human {gamma}D-crystallin.* Proceedings of the National Academy of Sciences, 2007. **104**(43): p. 16856-16861.
156. Capell, B.C. and F.S. Collins, *Human laminopathies: nuclei gone genetically awry.* Nat Rev Genet, 2006. **7**(12): p. 940-52.
157. Ackbarow, T. and M.J. Buehler, *Hierarchical Coexistence of Universality and Diversity Controls Robustness and Multi-Functionality in Protein Materials.* Journal of Computational and Theoretical Nanoscience, 2008. **5**: p. 1193-1204.
158. Kaminska, A., et al., *Small deletions disturb desmin architecture leading to breakdown of muscle cells and development of skeletal or cardioskeletal myopathy.* Human Genetics, 2004. **114**(3): p. 306-313.
159. Bryson, J.W., et al., *Protein design - a hierarachical approach.* Science, 1995. **270**(5238): p. 935-941.
160. Kirshenbaum, K., R.N. Zuckermann, and K.A. Dill, *Designing polymers that mimic biomolecules.* Current Opinion in Structural Biology, 1999. **9**(4): p. 530-535.
161. Ball, P., *Synthetic biology for nanotechnology.* Nanotechnology, 2005. **16**(1): p. R1-R8.
162. Lu, H. and K. Schulten, *Steered molecular dynamics simulations of force-induced protein domain unfolding.* Proteins-Structure Function and Genetics, 1999. **35**(4): p. 453-463.
163. Sotomayor, M. and K. Schulten, *Single-Molecule Experiments in Vitro and in Silico.* Science, 2007. **316**(5828): p. 1144-1148.
164. Gao, M., H. Lu, and K. Schulten, *Unfolding of titin domains studied by molecular dynamics simulations.* Journal of Muscle Research and Cell Motility, 2002. **23**(5-6): p. 513-521.
165. Boudko, S.P., et al., *Design and crystal structure of bacteriophage T4 mini-fibritin NCCF.* Journal of Molecular Biology, 2004. **339**(4): p. 927-935.
166. Warshel, A. and A. Papazyan, *Energy considerations show that low-barrier hydrogen bonds do not offer a catalytic advantage over ordinary hydrogen bonds.* PNAS, 1996. **93**(24): p. 13665-13670.
167. Lantz, M.A., et al., *Stretching the alpha-helix: a direct measure of the hydrogen-bond energy of a single-peptide molecule.* Chemical Physics Letters, 1999. **315**(1-2): p. 61-68.
168. Kageshima, M., et al., *Insight into conformational changes of a single alpha-helix peptide molecule through stiffness measurements.* Chemical Physics Letters, 2001. **343**(1-2): p. 77-82.
169. Griffith, A.A., *The phenomenon of rupture and flows in solids.* Phil. Trans. Roy. Soc. A, 1920. **221**: p. 163-198.
170. Rief, M., et al., *Single molecule force spectroscopy of spectrin repeats: Low unfolding forces in helix bundles.* Journal of Molecular Biology, 1999. **286**(2): p. 553-561.

171. Law, R., et al., *Influence of lateral association on forced unfolding of antiparallel spectrin heterodimers.* Journal of Biological Chemistry, 2004. **279**(16): p. 16410-16416.
172. Lenne, P.F., et al., *Stales and transitions during forced unfolding of a single spectrin repeat.* Febs Letters, 2000. **476**(3): p. 124-128.
173. Law, R., et al., *Cooperativity in forced unfolding of tandem spectrin repeats.* Biophysical Journal, 2003. **84**(1): p. 533-544.
174. Law, R., et al., *Pathway shifts and thermal softening in temperature-coupled forced unfolding of spectrin domains.* Biophysical Journal, 2003. **85**(5): p. 3286-3293.
175. Luhrs, T., et al., *3D structure of Alzheimer's amyloid-beta(1-42) fibrils.* Proceedings of the National Academy of Sciences of the United States of America, 2005. **102**(48): p. 17342-17347.
176. Mostaert, A.S., et al., *Nanoscale mechanical characterisation of amyloid fibrils discovered in a natural adhesive.* Journal Of Biological Physics, 2006. **32**(5): p. 393-401.
177. Smith, J.F., et al., *Characterization of the nanoscale properties of individual amyloid fibrils.* Proceedings of the National Academy of Sciences of the United States of America, 2006. **103**(43): p. 15806-15811.
178. Lezon, T.R., J.R. Banavar, and A. Maritan, *The origami of life.* Journal of Physics-Condensed Matter, 2006. **18**(3): p. 847-888.
179. Kitano, H., *Computational systems biology.* Nature, 2002. **420**(6912): p. 206-210.
180. Kitano, H., *Systems biology: A brief overview.* Science, 2002. **295**(5560): p. 1662-1664.
181. Chen, J.C.H., P.P. Chong, and Y.S. Chen, *Decision criteria consolidation: A theoretical foundation of Pareto Principle to Porter's Competitive Forces.* Journal of Organizational Computing and Electronic Commerce, 2001. **11**(1): p. 1-14.
182. Currey, J.D., *Bones: Structure and Mechanics.* 2002, Princeton, NJ: Princeton University Press.
183. Barthelat, F., et al., *On the mechanics of mother-of-pearl: A key feature in the material hierarchical structure.* Journal Of The Mechanics And Physics Of Solids, 2007. **55**(2): p. 306-337.
184. Heslot, H., *Artificial fibrous proteins: A review.* Biochimie, 1998. **80**(1): p. 19-31.
185. Ball, P., *Made to Measure: New Materials for the 21st Century.* 1997, Princeton, N.J., USA: Princeton University Press.
186. Buehler, M.J., *Rupture mechanics of vimentin intermediate filament tetramers.* Journal of Engineering Mechanics (ASCE), in submission.
187. Sokolova, A.V., et al., *Monitoring intermediate filament assembly by small-angle x-ray scattering reveals the molecular architecture of assembly intermediates.* Proceedings of the National Academy of Sciences of the United States of America, 2006. **103**(44): p. 16206-16211.
188. Buehler, M.J., *Rupture mechanics of vimentin intermediate filament tetramers.* Journal of Engineering Mechanics, 2008. **in press**.
189. Ackbarow, T. and M.J. Buehler, *Nanopatterned protein domains unify strength and robustness through hierarchical structures.* in submission.
190. Cui, X.Q., et al., *Biocatalytic generation of ppy-enzyme-CNT nanocomposite: From network assembly to film growth.* Journal of Physical Chemistry C, 2007. **111**(5): p. 2025-2031.

191. Hule, R., A., Pochan, D., J.,, *Polymer Nanocomposites for Biomedical Application.* MRS Bulletin, 2007. **32**(4): p. 5.
192. Winey, K.I., Vaia R.A.,, *Polymer Nanocomposites.* MRS Bulletin, 2007. **32**(4): p. 5.
193. Rho, J.Y., L. Kuhn-Spearing, and P. Zioupos, *Mechanical properties and the hierarchical structure of bone.* Medical Engineering and Physics, 1998. **20**(2): p. 92-102.
194. Currey, J.D., *Mechanical Properties of Mother of Pearl in Tension.* Proceedings of the Royal Society of London. Series B, Biological Sciences, 1977. **196**(1125): p. 443-463.
195. Menig, R., et al., *Quasi-static and dynamic mechanical response of Strombus gigas (conch) shells.* Materials Science & Engineering A, 2001. **297**(1-2): p. 203-211.
196. Tesch, W., et al., *Graded Microstructure and Mechanical Properties of Human Crown Dentin.* Calcified Tissue International, 2001. **69**(3): p. 147-157.
197. Courtney, T.H., *Mechanical behavior of materials.* 1990, New York, NY, USA: McGraw-Hill.
198. Sivaramakrishnan, S., et al., *Micromechanical properties of keratin intermediate filament networks.* Proceedings of the National Academy of Sciences of the United States of America, 2008. **105**: p. 889-894.
199. Buehler, M.J., S. Keten, and T. Ackbarow, *Theoretical and computational hierarchical nanomechanics of protein materials: Deformation and fracture.* Progress in Materials Science, 2008. **53**(8): p. 1101-1241.
200. Ackbarow, T., S. Keten, and M.J. Buehler, *Multi-time scale strength model of alpha-helical protein domains.* Journal of Physics: Condensed Matter, 2009. **21**: p. 035111.
201. Buehler, M.J. and H. Gao, *Dynamical fracture instabilities due to local hyperelasticity at crack tips* Nature, 2006. **439**: p. 307-310.
202. Grandbois, M., et al., *How strong is a covalent bond?* Science, 1999. **283**(5408): p. 1727-1730.
203. Plimpton, S., *Fast parallel algorithms for short-range molecular-dynamics* Journal of Computational Physics, 1995. **117**(1): p. 1-19.
204. Buehler, M.J., et al., *Threshold crack speed controls dynamical fracture of silicon single crystals.* Physical Review Letters, 2007. **99**.
205. Ackbarow, T., et al., *Hierarchies, multiple energy barriers and robustness govern the fracture mechanics of alpha-helical and beta-sheet protein domains.* P. Natl. Acad. Sci. USA, 2007. **104**: p. 16410-16415
206. Keten, S. and M.J. Buehler, *Geometric Confinement Governs the Rupture Strength of H-bond Assemblies at a Critical Length Scale.* Nano Letters, 2008. **8**(2): p. 743 - 748.
207. Hui, C.Y., *Crack blunting and the strength of soft elastic solids.* Proceedings of the Royal Society A: Mathematical, Physical and Engineering Sciences, 2003. **459**(2034): p. 1489-1516.
208. Lawn, B.R., *Fracture of brittle solids.* 1993: Cambridge University Press.
209. Gent, A.N. and C. Wang, *Fracture mechanics and cavitation in rubber-like solids.* journal of Materials Science, 1991. **26**(12): p. 3392-3395.
210. Pugno, N.M. and R.S. Ruoff, *Quantized fracture mechanics.* Philosophical Magazine, 2004. **84**(27): p. 2829-2845.

211. Buehler, M.J., *Rupture mechanics of vimentin intermediate filament tetramers.* Journal of Engineering Mechanics (ASCE), in press (accepted).
212. Herrmann, H., et al., *Intermediate filaments: from cell architecture to nanomechanics.* Nature Reviews Molecular Cell Biology, 2007. **8**(7): p. 562-573.
213. Fantner, G.E., et al., *Nanoscale ion mediated networks in bone: Osteopontin can repeatedly dissipate large amounts of energy.* Nano Letters, 2007. **7**(8): p. 2491-2498.
214. Fantner, G.E., et al., *Sacrificial bonds and hidden length: Unraveling molecular mesostructures in tough materials.* Biophysical Journal, 2006. **90**(4): p. 1411-1418.
215. Currey, J., *Biomaterials - Sacrificial bonds heal bone.* Nature, 2001. **414**(6865): p. 699-699.
216. Thompson, J.B., et al., *Bone indentation recovery time correlates with bond reforming time.* Nature, 2001. **414**(6865): p. 773-776.
217. Gao, H., et al., *Materials become insensitive to flaws at nanoscale: Lessons from nature.* P. Natl. Acad. Sci. USA, 2003. **100**(10): p. 5597-5600.
218. Bruck, H.A., J.J. Evans, and M.L. Peterson, *The role of mechanics in biological and biologically inspired materials.* Experimental Mechanics, 2002. **42**(4): p. 361-371.
219. Doyle, J., *Rules of engagement.* Nature, 2007. **446**: p. 860.
220. Csete, M.E. and J.C. Doyle, *Reverse engineering of biological complexity.* Science, 2002. **295**(5560): p. 1664-1669.
221. Alon, U., *Simplicity in biology.* Nature, 2007. **446**(7135): p. 497-497.
222. Baker, D., *A surprising simplicity to protein folding.* Nature, 2000. **405**(6782): p. 39-42.
223. Carlson, J.M. and J. Doyle, *Complexity and robustness.* Proceedings of the National Academy of Sciences of the United States of America, 2002. **99**: p. 2538-2545.
224. Carlson, J.M. and J. Doyle, *Highly optimized tolerance: A mechanism for power laws in designed systems.* Physical Review E, 1999. **60**(2): p. 1412-1427.
225. Alon, U., et al., *Robustness in bacterial chemotaxis.* Nature, 1999. **397**(6715): p. 168-171.
226. von Dassow, G., et al., *The segment polarity network is a robust development module.* Nature, 2000. **406**(6792): p. 188-192.
227. Wolfram, S., *A New Kind of Science.* 2002, Champaign: Wolfram Media.
228. Goel, N.S., Thompson, R.L., *Computer Simulations of Self-Organization in Biological Systems.* 1988, London & Sydney: Croom Helm.
229. Holland, J.H., *Hidden Order - How Adaptation Builds Complexity.* 1995, Reading, MA: Helix Books.
230. Case, J., Chilver, L., Ross C.T.F., *Strength of Materials and Structures.* 4th ed. 1999, New York: Arnold
231. Buehler, M., J., *Molecular nanomechanics of nascent bone: fibrillar toughening by mineralization.* Nanotechnology, 2007. **18**(29): p. 295102.
232. Berger, S.L., *Gene regulation - Local or global?* Nature, 2000. **408**(6811): p. 412-415.
233. Doyle, J.C., et al., *The "robust yet fragile" nature of the Internet.* Proceedings of the National Academy of Sciences of the United States of America, 2005. **102**(41): p. 14497-14502.
234. Doyle, J. and M. Csete, *Rules of engagement.* Nature, 2007. **446**(7138): p. 860.

235. Mucke, N., et al., *Investigation of the morphology of intermediate filaments adsorbed to different solid supports.* Journal of Structural Biology, 2005. **150**(3): p. 268-276.
236. Coulombe, P.A. and M.B. Omary, *'Hard' and 'soft' principles defining the structure, function and regulation of keratin intermediate filaments.* Current Opinion in Cell Biology, 2002. **14**(1): p. 110-122.
237. Gu, L.H. and P.A. Coulombe, *Keratin finction in skin epithelia: a broadening palette with surprising shades.* Current Opinion in Cell Biology, 2007. **19**(1): p. 13-23.
238. Kim, S., P. Wong, and P.A. Coulombe, *A keratin cytoskeletal protein regulates protein synthesis and epithelial cell growth.* Nature, 2006. **441**(7091): p. 362-365.
239. Zahn, H., et al., *Wool as a biological composite structure* Industrial & Engineering Chemistry Product Research and Development, 1980. **19**(4): p. 496-501.
240. Kreplak, L., et al., *New aspects of the alpha-helix to beta-sheet transition in stretched hard alpha-keratin fibers.* Biophysical Journal, 2004. **87**(1): p. 640-647.
241. Kreplak, L., et al., *A new deformation model of hard alpha-keratin fibers at the nanometer scale: Implications for hard alpha-keratin intermediate filament mechanical properties.* Biophysical Journal, 2002. **82**(4): p. 2265-2274.
242. Lorenzo Alibardi, M.T., *Immunological characterization of a newly developed antibody for localization of a beta-keratin in turtle epidermis.* Journal of Experimental Zoology Part B: Molecular and Developmental Evolution, 2007. **308B**(2): p. 200-208.
243. Dahl, K.N., D.E. Discher, and K.L. Wilson, *The nuclear envelope lamina network has elasticity and a compressibility limit suggestive of a "molecular shock absorber".* Molecular Biology of the Cell, 2004. **15**: p. 119A-120A.
244. Rowat, A.C., J. Lammerding, and J.H. Ipsen, *Mechanical properties of the cell nucleus and the effect of emerin deficiency.* Biophysical Journal, 2006. **91**(12): p. 4649-4664.
245. Mounkes, L., et al., *The laminopathies: nuclear structure meets disease.* Current Opinion in Genetics & Development, 2003. **13**(3): p. 223-230.
246. Olson, G.B., *Computational design of hierarchically structured materials.* Science, 1997. **277**(5330): p. 1237-1242.
247. Smeenk, J.M., et al., *Controlled assembly of macromolecular beta-sheet fibrils.* Angewandte Chemie-International Edition, 2005. **44**(13): p. 1968-1971.
248. Zhao, X.J. and S.G. Zhang, *Designer self-assembling peptide materials.* Macromolecular Bioscience, 2007. **7**(1): p. 13-22.
249. Zhao, X.J. and S.G. Zhang, *Molecular designer self-assembling peptides.* Chemical Society Reviews, 2006. **35**(11): p. 1105-1110.
250. Mershin, A., et al., *A classic assembly of nanobiomaterials.* Nature Biotechnology, 2005. **23**(11): p. 1379-1380.
251. Fraser, P., Bickmore, W., *Nuclear organization of the genome and the potential for gene regulation.* Nature, 2007. **447**(7143): p. 413-417.
252. Gao, H.J., *Application of fracture mechanics concepts to hierarchical biomechanics of bone and bone-like materials.* International Journal Of Fracture, 2006. **138**(1-4): p. 101-137.
253. Ji, B.H. and H.J. Gao, *Mechanical properties of nanostructure of biological materials.* Journal Of The Mechanics And Physics Of Solids, 2004. **52**(9): p. 1963-1990.

254. Gupta, H.S., et al., *Cooperative deformation of mineral and collagen in bone at the nanoscale.* P. Natl. Acad. Sci. USA, 2006. **103**: p. 17741-17746.
255. Webb, R., *Social science: the urban organism.* Nature, 2007. **446**(7138): p. 869.
256. Ferguson, N., *Capturing human behaviour.* Nature, 2007. **446**(7137): p. 733-733.
257. Mandelbrot, B.B., *Fractal geometry - what it is, and what does it do.* Proceedings of the Royal Society of London Series a-Mathematical Physical and Engineering Sciences, 1989. **423**(1864): p. 3-16.
258. Alexander, R.M., *The gecko's foot. Bio-inspiration: Engineered from nature.* Nature, 2005. **438**(7065): p. 166-166.

I want morebooks!

Buy your books fast and straightforward online - at one of world's fastest growing online book stores! Environmentally sound due to Print-on-Demand technologies.

Buy your books online at
www.morebooks.shop

Kaufen Sie Ihre Bücher schnell und unkompliziert online – auf einer der am schnellsten wachsenden Buchhandelsplattformen weltweit! Dank Print-On-Demand umwelt- und ressourcenschonend produziert.

Bücher schneller online kaufen
www.morebooks.shop

KS OmniScriptum Publishing
Brivibas gatve 197
LV-1039 Riga, Latvia
Telefax: +371 686 204 55

info@omniscriptum.com
www.omniscriptum.com

Printed by Books on Demand GmbH, Norderstedt / Germany